SpringerBriefs in Optimization

Series Editors

Panos M. Pardalos
János D. Pintér
Stephen M. Robinson
Tamás Terlaky
My T. Thai

SpringerBriefs in Optimization showcases algorithmic and theoretical techniques, case studies, and applications within the broad-based field of optimization. Manuscripts related to the ever-growing applications of optimization in applied mathematics, engineering, medicine, economics, and other applied sciences are encouraged.

For further volumes:
http://www.springer.com/series/8918

Boris Goldengorin • Panos M. Pardalos

Data Correcting Approaches
in Combinatorial
Optimization

 Springer

Boris Goldengorin
Laboratory of Algorithms and Technologies
  for Networks Analysis (LATNA)
  and Department of Higher Mathematics
National Research University
Higher School of Economics, Russia

and

Operations Department
University of Groningen
The Netherlands

Panos M. Pardalos
Center for Applied Optimization
Department of Industrial
  and Systems Engineering
401 Weil Hall
University of Florida
Gainesville, FL, USA

and

Laboratory of Algorithms and Technologies
  for Networks Analysis (LATNA)
National Research University
Higher School of Economics
Moscow, Russia

ISSN 2190-8354   ISSN 2191-575X (electronic)
ISBN 978-1-4614-5285-0   ISBN 978-1-4614-5286-7 (eBook)
DOI 10.1007/978-1-4614-5286-7
Springer New York Heidelberg Dordrecht London

Library of Congress Control Number: 2012947867

Mathematics Subject Classification (2010): 90-02, 90C27, 90C09, 90C10, 49L20, 90C35, 90B06, 90B10, 90B40, 90B80, 68R10

Printed on acid-free paper

Springer is part of Springer Science+Business Media (www.springer.com)

Preface

The goal of this book is to develop methodological principles of data correcting (DC) algorithms for solving NP-hard problems in combinatorial optimization. We consider two large classes of NP-hard problems defined either on the set of all subsets of a finite set (see the maximization of submodular functions, quadratic cost partition, and simple plant location problems) or on the set of all permutations of a finite set (see the Traveling Salesman Problem, TSP).

This book is organized as follows: in Chap. 1 we motivate the DC approach by its application to a single real-valued function defined on a continuous domain, which has a finite range, and describe how this approach might be applied to a general combinatorial optimization problem including its implementation for the asymmetric TSP (ATSP).

The first purpose of Chap. 2 is to make more accessible to the Western community some long-standing theoretical results about the structure of local and global maxima of submodular functions due to [29, 89] including Cherenin's excluding rules and his dichotomy algorithm (see [27, 28, 118]). We use Cherenin's dichotomy algorithm for determining a polynomially solvable class of submodular functions (*PP-functions*) and show that PP-functions contain precisely one component of strict local maxima. The second purpose of Chap. 2 is to present a generalization of Cherenin's excluding rules. This result is a base of DC algorithms for the maximization (minimization) of submodular (supermodular) functions presented in Chap. 3.

In Chap. 3 we present the DC algorithm for maximization of submodular functions; it is a recursive Branch-and-Bound (BnB)-type algorithm (see e.g., [6]). In the DC algorithm the values of a given submodular function are "heuristically" corrected at each branching step in such a way that the new (corrected) submodular function will be as close as possible to a polynomially solvable instance from the class of submodular PP-functions (instances), and the result satisfies a prescribed accuracy parameter. The working of the DC algorithm is illustrated by means of an instance of the simple plant location problem (SPLP). Computational results, obtained for the quadratic cost partition (QCP) problem, show that the computing results of the DC algorithm in general are better than the computational results known in the current literature (see, e.g., [8, 55, 102, 114, 115, 119, 120]), not only for sparse graphs but also for nonsparse graphs (with density more than 40 %) often with speeds 100 times faster. We further improve the DC algorithm for submodular functions by introducing an extended PP-function. Our computational experiments with the improved DC algorithm on QCP instances, similar to those in [102], allow us to solve QCP instances on dense graphs with number of vertices up to 500 within 10 min on a standard personal computer.

In Chap. 4 we deal with a pseudo-Boolean representation of the SPLP (see, e.g., [13, 38, 80]).

We improve the class of Branch and Peg algorithms (see [67, 68]) by using SPLP-specific bounds (suggested in [44]) and preprocessing rules (coined in [91]) in Chap. 4. We further incorporate a new reduction procedure based on data correcting, which is stronger than the preprocessing rules from [91], to reduce the original instance to a smaller "core" instance, and then solve it using a procedure based on DC algorithm developed in Chap. 3. Computational experiments with the DC algorithm adapted to the SPLP on benchmark instances suggest that the algorithm compares well with other algorithms known for the SPLP (see [69]).

In the summary of this book we discuss future research directions for DC approach based on the main results presented in the conclusions of the chapters.

Nizhny Novgorod, Russia   Boris Goldengorin
Groningen, The Netherlands
Gainsville, FL   Panos M. Pardalos

Acknowledgements

The research reported in this book was started many years ago at the Computer Science Department of the Kazakh State University and was continued at the Department of Econometrics and Operations Research of the University of Groningen.

This book could not have been written without the greatest access to the e-Library of University of Groningen, The Netherlands granted by the Faculty of Economics. Boris' applications in 2011 and 2012 to have an access to the e-Library were supported by the former Head of Operations Department Maarten van der Vlerk and Vice-Dean of Faculty of Economics Teun van Duinen, respectively. Dear Maarten and Teun, your support is unforgettable, thanks a lot!

The preparation of this book is started at the Laboratory of Algorithms and Technologies for Networks Analysis (LATNA) recently established within the National Research University Higher School of Economics (NRU HSE) in Nizhny Novgorod, Russian Federation (RF) under the great supervision of the coauthor of this book Panos M. Pardalos. The financial support by LATNA Laboratory, NRU HSE, RF government grant ag. 11.G34.31.0057 is gratefully acknowledged.

It is a great pleasure for us to thank all the persons who contributed to the accomplishment of the book. Eduard Babkin encouraged Boris to move from Groningen (The Netherlands) to Nizhny Novgorod and Valery Kalyagin helped in many ways that Boris movement become a reality. A special thanks goes to Valery's wife—Tamara who cared about Boris's three children and his family, especially in the first couple of months after arriving Nizhny Novgorod. Your help, Tamara, is very appreciated!

In the recent many years we have had a pleasure to discuss different fragments of the data correcting approach with many colleagues and friends. The first important support was provided by the Nobel Prize winner in economics, Leonid Vitaliyevich Kantorovich, who has recommended the paper [59] to be published in the most prestigious scientific outlet *Soviet Math. Doklady*.

Boris is very thankful to the late Gert A. Tijssen. Doing joint research, writing articles together, and discussing the results of computational experiments during many years have been very pleasant and fruitful. Gert's insightful comments on the nature of DC approach had led to the following publications [66, 69].

This book has benefited from many discussions and comments; corrections have been provided at different stages by the following colleagues S. Benati, Z. Bolotin, M. Ebben, D. Ghosh, G. Gutin, W.K. Klein Haneveld, J.W. Nieuwenhuis, G. Sierksma, F. Tardella, and M. Turkensteen.

This book is dedicated to Boris' and Panos' loving families: Ljana, Polina, Nicolai, Vitaliy, Mark and Rosemary, Miltiades, respectively.

Contents

Chapter 1
Introduction

Combinatorial optimization problems are those where one has to choose among a countable number of alternatives. Managerial applications of such problems are often concerned with the efficient allocation of limited resources to meet desired objectives, for example, increasing productivity when the set of solutions (variants) is finite. Constraints on basic resources, such as labor, facilities, supplies, or capital, restrict the possible alternatives to those that are considered *feasible*. Applications of these problems include goods distribution, production scheduling, capital budgeting, facility location, the design of communication and transportation networks, the design of very large scale integration (VLSI) circuits, the design of automated production systems, artificial intelligence, machine learning, and software engineering.

In mathematics there are applications to the subject of combinatorics, graph theory, probability theory, auction theory, and logic.

Statistical applications include problems of data analysis and reliability. The number and variety of applications of combinatorial optimization models are so great that we only can provide references for some of them (see, e.g., [40, 94]).

Combinatorial optimization is a part of mathematical optimization that is related to operations research, algorithm theory, and computational complexity theory. It has important applications in several fields, including artificial intelligence, machine learning, mathematics, auction theory, energy, biomedicine, and software engineering.

The combinatorial nature of the above-mentioned problems arises from the fact that in many real-world problems, activities and resources, for instance machines and people, are indivisible. Also, many problems have only a finite number of alternative choices and consequently can be formulated as combinatorial optimization problems—the word "combinatorial" refers to the fact that a feasible solution to a combinatorial optimization problem can be constructed as a combination of indivisible objects. It is relatively easy to construct an algorithm which computes the cost of each feasible solution and keeps the best in mind. Unfortunately, such an exhaustive enumeration algorithm is usually impractical when there are more

B. Goldengorin and P.M. Pardalos, *Data Correcting Approaches in Combinatorial Optimization*, SpringerBriefs in Optimization, DOI 10.1007/978-1-4614-5286-7_1,
© Boris Goldengorin, Panos M. Pardalos 2012

than 20 objects, since there are simply too many feasible solutions. For example, the Traveling Salesman Problem (TSP) defined on the set of n cities has $(n-1)!$ different feasible solutions (tours). Even assuming that we have a very fast computer that can evaluate one million tours per second, and we have 20 cities, an enumerative algorithm would take over 750 centuries to evaluate all possible tours (see, e.g., [41]).

Computer scientists have found that certain types of problems, called NP-hard problems, are intractable (see, e.g., [48]). Roughly speaking this means that the time it takes to solve any typical NP-hard problem seems to grow exponentially as the amount of input data (instance) increases (see, e.g., [37, 111]). On the other hand, for many NP-hard problems we can provide provable analytic or algorithmic characterizations of the instances input data that guarantee a polynomial time solution algorithm for the corresponding instances. These instances are called *polynomially solvable special cases* of the combinatorial optimization problem (see, e.g., [22]).

Polynomially solvable special cases of combinatorial optimization problems have long been studied in the literature (see, e.g., for the TSP [52, 85]). Apart from being mathematical curiosities, they often provide important insights for serious problem solving. In fact, the concluding paragraph of [52] states the following, regarding polynomially solvable special cases for the TSP.

> " ⋯ We believe, however, that in the long run the greatest importance of these special cases
> will be for approximation algorithms. Much remains to be done in this area."

This book is a step in the direction of incorporating polynomially solvable special cases into approximation and exact algorithms (see [53]). We propose a *data correcting (DC) algorithm*—an approximation algorithm that makes use of polynomially solvable special cases to arrive at high-quality solutions. The basic insight that leads to this algorithm is the fact that it is often easy to compute an upper bound on the difference in cost between an optimal solution of a problem instance and any feasible solution to the instance. The results obtained with this algorithm are very promising (see the computational results in the corresponding sections of this book).

The approximation in the DC algorithm is in terms of an *accuracy parameter*, which is an upper bound on the difference between the objective value of an optimal solution to the instance and that of a solution returned by the DC algorithm. Note that this is not expressed as a fraction of the optimal objective value for this instance. In this respect, the algorithm is different from common ε-optimal algorithms, in which ε is defined as a fraction of the optimal objective function value.

Even though the algorithm is meant mainly for NP-hard combinatorial optimization problems, it can be used for functions defined on a continuous domain too. We will, in fact, motivate the DC algorithm in the next section using a function defined on a continuous domain that has a finite range. We then show in Sect. 1.2, how this approach can be adapted for NP-hard optimization problems, using the Asymmetric TSP (ATSP) as an illustration. We conclude the introductory chapter with a summary of the remaining chapters of this book.

1.1 Data Correcting Approach for Real-Valued Functions

Consider a real-valued function $f : \mathscr{D} \to \mathfrak{R}$, where $\mathscr{D}$ is the domain on which the function is defined. Here we discuss a minimization problem; the maximization version of the problem can be dealt with in a similar manner (see [53]). We assume that function values of f are easy to compute, but finding the minimum of f over a subdomain takes more than a reasonable amount of computing time. We concern ourselves with the problem of finding α-minimal solutions to the function f over $\mathscr{D}$, i.e., the problem of finding a member of $\{x \,|\, x \in \mathscr{D}, f(x) \leq f(x^\star) + \alpha\}$, where $x^\star \in \arg\min_{x \in \mathscr{D}}\{f(x)\}$, and α is a predefined *accuracy parameter*.

Let us assume that $\{\mathscr{D}_1, \ldots, \mathscr{D}_p\}$ is a partition of the domain $\mathscr{D}$. Let us further assume that for each of the sub-domains $\mathscr{D}_i$ of $\mathscr{D}$, we are able to find functions $g_i : \mathscr{D}_i \to \mathfrak{R}$, which are easy to minimize over $\mathscr{D}_i$, and such that

$$|f(x) - g_i(x)| \leq \frac{\alpha}{2} \quad \forall x \in \mathscr{D}_i. \tag{1.1}$$

We call such easily minimizable functions *regular* (see [59]).

Theorem 1.1 provides an interesting approximation $f(x^\alpha)$ of the unknown global optimum $f(x^\star)$, where $f(x^\alpha)$ is the minimum of all values $f(x_1^\alpha), \ldots, f(x_p^\alpha)$ and x_i^α is the minimum of a function g_i on $\mathscr{D}_i$ satisfying (1.1).

Theorem 1.1. *For $i = 1, \ldots, p$, let $x_i^\alpha \in \arg\min_{x \in \mathscr{D}_i}\{g_i(x)\}$, and $x^\alpha \in \arg\min_i\{f(x_i^\alpha)\}$. Moreover, let $x^\star \in \arg\min_{x \in \mathscr{D}}\{f(x)\}$. Then*

$$f(x^\alpha) \leq f(x^\star) + \alpha.$$

Proof. Let $x_i^\star \in \arg\min_{x \in \mathscr{D}_i}\{f(x)\}$. Then for $i = 1, \ldots, p$, $f(x_i^\alpha) - \frac{\alpha}{2} \leq g_i(x_i^\alpha) \leq g_i(x_i^\star) \leq f(x_i^\star) + \frac{\alpha}{2}$, i.e., $f(x_i^\alpha) \leq f(x_i^\star) + \alpha$. Thus $\min_i\{f(x_i^\alpha)\} \leq \min_i\{f(x_i^\star)\} + \alpha$, which proves the result. □

Notice that $x^\star$ and x^α do not need to be in the same sub-domain of $\mathscr{D}$. Theorem 1.1 forms the basis of the DC algorithm to find an approximate minimum of a function f over a certain domain $\mathscr{D}$. The procedure consists of three steps. In the first step the domain $\mathscr{D}$ of the function is partitioned into several sub-domains. In the second step f is approximated in each of the sub-domains by regular functions satisfying the condition in expression (1.1) and a minimum point of the regular function is obtained. Finally, the third step, in which the minimum points computed in the second step are considered and the best among them is chosen as the output. This procedure can be further strengthened by using lower bounds to check if a given sub-domain can possibly lead to a solution better than any found thus far. The approximation of f by regular functions g_i is called *data correcting*, since an easy way of obtaining the regular functions is by altering the data that describe f. A pseudocode of the algorithm, which we call *Procedure DC*, is provided in Fig. 1.1.

Procedure DC
Input: f, $\mathcal{D}$, α.
Output: $x^\alpha \in \mathcal{D}$ such that $f(x^\alpha) \leq \min\{f(x)|x \in \mathcal{D}\} + \alpha$.
Code:

```
 1 begin
 2        bestvalue := ∞;
 3        create a partition {𝒟₁,...,𝒟ₙ} of 𝒟;
 4        for each sub-domain 𝒟ᵢ
 5        begin
 6              fᵢ := a lower bound to f(x), x ∈ 𝒟ᵢ;
 7              if fᵢ ≥ bestvalue
 8              then fathom 𝒟ᵢ
 9              else
10                    if the constructed regular function gᵢ(x)
11                        satisfies (1.1)
12                    then calculate xᵢᵅ ∈ arg minₓ∈𝒟ᵢ{gᵢ(x)}; and
13                          go to line 15
14                    else go to line 3;
15              if f(xᵢᵅ) < bestvalue;
16              then begin
17                    xᵅ := xᵢᵅ;
18                    bestvalue := f(xᵢᵅ);
19                    end;
20        end;
21        return xᵅ;
22 end.
```

Fig. 1.1 A DC algorithm for a real-valued function

Lines 6–8 in the code carry out the bounding process, and lines 9 and 14 implement the process of computing the minimum of the regular function over this sub-domain. The code in lines 15–19 remembers the best minimum. By Theorem 1.1, the solution chosen by the code in lines 15–19 is an α-minimum of f, and therefore this solution is returned by the algorithm in line 21. The difference between this algorithm and the usual branch-and-bound (BnB) method is that in the DC-algorithm in each domain $\mathcal{D}_i$, the original function is approximated by a regular function, as can be seen in line 9. Note that a regular function should not be either a lower or an upper bound to the unknown optimal value on D.

We will now illustrate the DC algorithm through an example. The example that we choose is one of a real-valued function of one variable, since these are some of the simplest functions to visualize.

Consider the problem of finding an α-minimum of the function f shown in Fig. 1.2. The function is defined on the domain $\mathcal{D}$ and is assumed to be analytically intractable.

The DC approach can be used to solve the problem above, i.e., of finding a solution $x^\alpha \in D$ such that $f(x^\alpha) \leq \min\{f(x)|x \in \mathcal{D}\} + \alpha$.

Fig. 1.2 A general function f

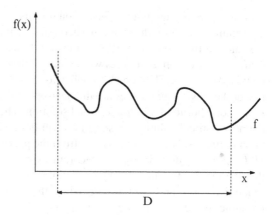

Fig. 1.3 Illustrating the DC approach on f

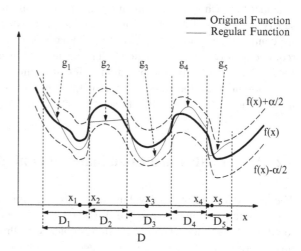

Consider the partition $\{D_1, D_2, D_3, D_4, D_5\}$ of D shown in Fig. 1.3. Let us suppose that we have a regular function $g_1(x)$ such that $|g_1(x) - f(x)| \leq \frac{\alpha}{2}, \forall x \in D_1$. Assume also that x_1 is a minimum point of $g_1(x)$ in D_1. Since this is the best solution that we have so far, we store x_1 as an α-minimal solution to $f(x)$ in the domain D_1. We then consider the next interval in the partition, D_2. We construct a regular function $g_2(x)$ with $|g_2(x) - f(x)| \leq \frac{\alpha}{2}, \forall x \in D_2$, and find x_2, its minimum point over D_2. Since $f(x_2) > f(x_1)$ (see Fig. 1.3), we retain x_1 as our α-optimal solution over $D_1 \cup D_2$. Proceeding in this manner, we examine $f(x)$ in D_3 through D_5, compute regular functions $g_3(x)$ through $g_5(x)$ for these domains, and compute x_3 through x_5. In this example, x_3 replaces x_1 as our α-minimal solution after consideration of D_3, and remains so until the end. At the end of the algorithm, x_3 is returned as a value of x^α.

There are four points worth noting at this stage. The first is that we need to examine all the sub-domains in the original domain before we return a near-optimal solution using this approach. The reason for this is very clear. The correctness of the algorithm depends on the result in Theorem 1.1, and this theorem only concerns the *best* among the minima of each of the sub-domains. For instance, in the previous example, if we stop as soon as we obtain the first α-optimal solution x_1 we would be mistaken, since Theorem 1.1 applies to x_1 only over $D_1 \cup D_2$. The second point is that there is no guarantee that the near-optimal solution returned by DC will be in the neighborhood of a true optimal solution. There is in fact, nothing preventing the near-optimal solution existing in a sub-domain different from the sub-domain of an optimal solution, as is evident from the previous example. The true minimum of f lies in the domain D_5, but DC returns x_3, which is in D_3. The third point is that the regular functions $g_i(x)$ approximating $f(x)$ do not need to have the same functional form, and are not induced by the form of $f(x)$ in that domain. In this Example, $g_1(x)$ is quadratic, while $g_2(x)$ is linear. In general it is not always the case that for each specific domain D_i and a fixed class of regular functions we can satisfy the required quality of approximation expressed by (1.1). In such a case we continue the splitting process of the domain D_i into sub-domains such that the required quality of approximation will be achieved. It is not necessarily that $\alpha_i = \alpha$ for $i = 1,\ldots,n$. In this case based on different α_i an optimal approximation value $\alpha = \gamma$ might be computed (see Theorem 3.2). Finally, for the proof of Theorem 1.1, it is sufficient for $\{\mathscr{D}_1,\ldots,\mathscr{D}_n\}$ to be a cover of $\mathscr{D}$ (as opposed to a partition).

1.2 DC for NP-Hard Combinatorial Optimization Problems

The DC methodology described in the previous section can be incorporated into an implicit enumeration scheme (like BnB) and used to obtain near-optimal solutions to NP-hard combinatorial optimization problems. In this section we describe how this incorporation is achieved for a general combinatorial optimization problem. A combinatorial optimization problem (COP) $(\mathscr{G}, C, \mathscr{S}, f_C)$ is the problem of finding

$$S^* \in \arg OPT\{f_C(S) \mid S \in \mathscr{S}\},$$

where $C : \mathscr{G} \to \mathfrak{R}$ is the instance of the problem with ground set $\mathscr{G}$ satisfying $|\mathscr{G}| = n\ (n \geq 1)$, $\mathscr{S} \subseteq 2^{\mathscr{G}}$ the set of feasible solutions, and $f_C : \mathscr{S} \to \mathfrak{R}$ the objective function. In this section it is assumed that $OPT = \min$, so that we only consider minimization problems. In case of the TSP (see, e.g., [78, 100]), i.e., the problem of finding a shortest tour visiting a given set $\{1,\ldots,m\}$ covering the m locations, $\mathscr{G}$ is the set of edges connecting these locations, $\mathscr{S}$ the set of possible tours (also called *Hamiltonian* tours) along the m locations, C is the $m \times m$ distance matrix on $\mathscr{G}$ with $n = m^2$, and $f_C(S) = \sum_{s \in S} c(s)$ with $C = \{c(s)\}$ for each $s \in S \in \mathscr{G}$ is the cost of the tour.

Different entries of C define different instances of the same COP. Let $(\mathscr{G}, C_1, \mathscr{S}, f_{C_1})$ and $(\mathscr{G}, C_2, \mathscr{S}, f_{C_2})$ define two instances C_1 and C_2 of the same COP. The function $\rho(\cdot,\cdot)$ is called a *proximity measure* for the COP $(\mathcal{G},C,\mathcal{S},f_C)$, if for each pair of instances $C_1 = \{c_1(s)\}$ and $C_2 = \{c_2(s)\}$ of this COP, it holds that

$$|f_{C_1}(S_1^*) - f_{C_2}(S_2^*)| \leq \rho(C_1,C_2), \tag{1.2}$$

for each $S_i^* \in \arg\min\{f_{C_i}(S) \mid S \in \mathcal{S}\}$ for $i = 1,2$. Here $s \in S$ and in case of the TSP s is an edge (arc) of a Hamiltonian cycle S.

Theorem 1.2. *Let $(\mathcal{G},C,\mathcal{S},f_C)$ be a COP with ground set $\mathcal{G}$.*
Then,

$$\rho_1(C_1,C_2) = \sum_{s \in \mathcal{G}} |c_1(s) - c_2(s)| \tag{1.3}$$

is a proximity measure, if either $f_C = \sum_c$, or $f_C = \max_c$, where for each $S \in \mathcal{S}$ it holds that $\sum_c(S) = \sum_{s \in S} c(s)$, and $\max_c(S) = \max_{s \in S} c(s)$.

Proof. Take any two instances C_1 and C_2 of the COP $(\mathcal{G},C,\mathcal{S},f_C)$, and any $S_i^* \in \arg\min\{f_{C_i}(S) \mid S \in \mathcal{S}\}$; $i = 1,2$. Assume that $f_{C_1}(S_1^*) - f_{C_2}(S_2^*) \geq 0$. Then,

$$|f_{C_1}(S_1^*) - f_{C_2}(S_2^*)| = f_{C_1}(S_1^*) - f_{C_2}(S_2^*) \leq f_{C_1}(S_2^*) - f_{C_2}(S_2^*)$$
$$= \sum_{s \in S_2^*} [c_1(s) - c_2(s)] \leq \sum_{s \in S_2^*} |c_1(s) - c_2(s)| \leq \sum_{s \in \mathcal{G}} |c_1(s) - c_2(s)|.$$

The other case can be shown along similar lines. □

Note that the complexity of computing the proximity measure $\rho_1(C_1,C_2)$ in Theorem 1.2 is polynomial, although, in general, that of computing $f_C(S^*)$ is not.

Corollary 1.1. *Let $(\mathcal{G},C_1,\mathcal{S},f_{C_1})$ and $(\mathcal{G},C_2,\mathcal{S},f_{C_2})$ be two instances C_1 and C_2 of the same COP with the corresponding proximity measure $\rho(C_1,C_2)$, optimal solutions S_1^* and S_2^*, respectively. There are two virtual bounds for $f_{C_1}(S_1^*)$ as follows:*

$$f_{C_2}(S_2^*) - \rho(C_1,C_2) \leq f_{C_1}(S_1^*), \tag{1.4}$$

$$f_{C_1}(S_1^*) \leq f_{C_2}(S_2^*) + \rho(C_1,C_2). \tag{1.5}$$

Proof. The proof is straightforward from (1.2). □

Both virtual bounds might be useful within a BnB type algorithm either to discard a subproblem by means of the lower virtual bound (1.4) or to improve the value of the best currently found feasible solution by means of a feasible solution related to the upper virtual bound (1.5). It means that tight proximity measures might be used as a new source of lower and upper virtual bounds in combinatorial optimization (see, e.g., [94]).

One way of implementing the DC step as formulated in Sect. 1.1 for COP $(\mathcal{G},C,\mathcal{S},f_C)$ is illustrated by means of the ATSP, i.e., the TSP in which the distance

c_{ij} from location i to j is not necessarily equal to the distance c_{ji} from j to i. In order to execute the DC step, we construct a polynomially solvable relaxation $(\mathscr{G}, C, \mathscr{S}_R, f_C)$ of the original problem $(\mathscr{G}, C, \mathscr{S}, f_C)$ with the optimal solution $S_R \in \mathscr{S}_R$, namely, the Assignment Problem (AP) defined on the same ground set $\mathscr{G}$ with the same distance matrix C, and the set of feasible solutions $\mathscr{S}_R$ being (not necessary cyclic) permutations of $1, \ldots, m$, i.e. $\mathscr{S}_R \supset \mathscr{S}$. It means that $S_R \in \mathscr{S}_R$ need not be feasible to $(\mathscr{G}, C, \mathscr{S}, f_C)$. We next construct a "good" solution S_F to $(\mathscr{G}, C, \mathscr{S}, f_C)$ based on S_R. To that end, we apply a patching operation on S_R to obtain the cyclic permutation (tour) S_F (see, e.g., [88]). We also construct an instance C_F for which $(\mathscr{G}, C_F, \mathscr{S}, f_C)$ will have S_F as an optimal solution. Note that, in general, $C_F \neq C$. C_F is called the *corrected instance* based on C. Clearly, the *corrected* COP $(\mathscr{G}, C_F, \mathscr{S}, f_C)$ is polynomially solvable. The proximity measure $\rho(C, C_F) = \sum_{i=1}^{m} \sum_{j=1}^{m} |c_{ij} - c_{ij}^F|$ then is an upper bound to the difference between the costs of S_F and of an optimal solution to $(\mathscr{G}, C, \mathscr{S}, f_C)$. The following example illustrates this technique.

Consider the 6-city ATSP instance with the distance matrix $C = [c_{ij}]$ shown below. Solving the AP on C by means of the Hungarian method (see, e.g., [97])

C	1	2	3	4	5	6
1	–	10	16	19	25	22
2	19	–	10	13	13	10
3	10	28	–	22	16	13
4	19	25	13	–	10	19
5	16	22	19	13	–	11
6	13	22	15	13	10	–

results in the following reduced distance matrix $C^H = [c_{ij}^H]$.

C^H	1	2	3	4	5	6
1	–	0	6	7	16	12
2	9	–	0	1	4	0
3	0	18	–	10	7	3
4	8	14	2	–	0	8
5	5	11	8	0	–	0
6	2	11	4	0	0	–

Recall that the Hungarian method is based on the following two observations. If we add (subtract) an arbitrary finite constant A to all entries of either row i or column j of the distance matrix C, then the set of feasible solutions to the AP is not changed. Hence, any distance matrix can be reduced into a distance matrix with nonnegative entries. Therefore, an optimal solution to the original AP can be represented by a set of n "independent" zeros in a reduced distance matrix. We call a set of n zeros *independent* if each pair of them is located in a set of pairwise distinct rows and

Fig. 1.4 A patching operation

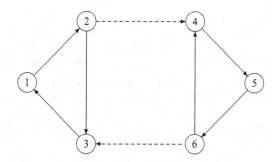

columns. Note that the value of an optimal solution to the original AP will differ from the optimal solution of the reduced AP by the sum of all added (subtracted) constants leading to the reduced matrix.

In this example, $S_R = \{(1231), (4564)\}$. Using patching techniques (see, [88]), we obtain the solution $S_F = (1245631)$. Here, the patching technique involves the deletion of an arc (i_1, j_1) from $(1, 231)$ and an arc (i_2, j_2) from $(4, 564)$, and insertion of the arcs (i_1, j_2) and (i_2, j_1) into the remaining set of arcs such that the value of $c_{i_1 j_2} + c_{i_2 j_1}$ is minimal. In this example $(i_1, j_1) = (2, 3)$ and $(i_2, j_2) = (6, 4)$ (see Fig. 1.4).

Notice that the patched solution (1245631) would be an optimal solution to the AP if c_{24}^H and c_{63}^H had been set to zero in C^H, and that would have been the case, if c_{24} and c_{63} had been initially reduced by 1 and 4, respectively, i.e., if the distance matrix in the original ATSP instance had been C_F, as shown below.

C_F	1	2	3	4	5	6
1	–	10	16	19	25	22
2	19	–	10	12	13	10
3	10	28	–	22	16	13
4	19	25	13	–	10	19
5	16	22	19	13	–	11
6	13	22	11	13	10	–

Therefore C_F is the corrected distance matrix. The proximity measure satisfies $\rho_1(C, C_F) = \sum_{i=1}^{6} \sum_{j=1}^{6} |c_{ij} - c_{ij}^F| = |c_{24} - c_{24}^F| + |c_{63} - c_{63}^F| = 4 + 1 = 5$.

Since a proximity measure is an upper bound for the difference between the optimal value of the original instance (in our example C) and the corrected instance (in our example C_F), it may be clear that the stronger the bound, the better would be the performance of enumeration algorithms depending on such bounds. In this sense, ρ_2, as defined below, is a stronger proximity measure for ATSP instances:

$$\rho_2(C, C_F) = \min \left\{ \sum_{i=1}^{m} \max_{1 \leq j \leq n} |c_{ij} - c_{ij}^F|, \sum_{j=1}^{m} \max_{1 \leq i \leq n} |c_{ij} - c_{ij}^F| \right\}. \tag{1.6}$$

Note that both measures, ρ_1 and ρ_2, use a set of corrected entries of C which are located in pairwise distinct rows and columns on each correcting step. Hence, the values of ρ_1 and ρ_2 will be the same and equal to the cost of patching the solution S_F for each correcting step. Consider the ATSP instance in the above example. The cost $c_{24} + c_{63} - c_{23}^F - c_{64}^F$ of patching the solution $\{(1231),(4564)\}$ to (1245631) is exactly equal to the values of $\rho_1(C,C_F)$ and $\rho_2(C,C_F)$. Actually, after two or more correcting steps, the set of corrected entries does not necessary contain entries located in pairwise distinct rows and columns. This means that the values of the proximity measures ρ_1 and ρ_2 become different, actually they will satisfy $\rho_2 < \rho_1$. It means that, using the proximity measure ρ_2, one can save execution time. In case of the ATSP, the values of ρ_1 and ρ_2 become available as by-products of computing the best patching at the *first* correcting step.

1.3   The DC Approach in Action

The correcting step in DC algorithms can be represented as the following *approximation problem*:

$$\min\{\rho(C,C_F)\,|\,C_F \in \mathscr{R}\} = \rho(C,C_F^0) \tag{1.7}$$

for any COP $(\mathscr{G},C,\mathscr{S},f_C)$. In (1.7), $\mathscr{R}$ is a set of regular or polynomially solvable instances of the same size as C. Actually, C_F^0 is an instance as close as possible to the given instance C, and $\rho(\cdot,\cdot)$ is a proximity measure as defined in the previous section. It is clear that the computational complexity of problem (1.7) depends both on the structure of the set $\mathscr{R}$ and the proximity measure $\rho(C,C_F)$. For many classes $\mathscr{R}$ the approximation problem (1.7) is NP-hard (see, e.g., [64]). In the DC approach we therefore just use a heuristic for solving (1.7), say with solution $C_F^a \in \mathscr{R}$, so that $\rho(C,C_F^0) \leq \rho(C,C_F^a)$. Recall that Theorem 1.2 enables us to decide "how far" the solution of $(\mathscr{G},C_F^a,\mathscr{S},f_{C_F^a})$ is "away" from $(\mathscr{G},C,\mathscr{S},f_C)$. Let α be the accuracy parameter. In fact, if $\rho(C,C_F^a) \leq \alpha$, we have found an α-optimal solution to the original problem with cost vector C, so we are done. If $\rho(C,C_F^a) > \alpha$, we partition (branch) the set $\mathscr{S}$ of feasible solutions into a number of new sets. Like in usual BnB, the partition is obtained with so-called branching rules. These rules are problem specific. However, in case of DC, it turns out that a branching rule based on *upper tolerances* (being bounds on the values of the input parameters, whereas within these bounds the same optimal solution holds; see e.g., [131]) on the entries in the currently best solution is the best choice for the given $\mathscr{R}$ (see [39]). This fact will not be elaborated here (see [70]). More details about upper tolerances-based branching rules might be found in [51, 130]. Figure 1.5 presents the pseudocode of a recursive version of the DC method for combinatorial optimization problems with minimization objective. The input is the cost vector C, with its size n, the feasible solutions set $\mathscr{S}$, and the accuracy parameter α. Notice that lines 2 and 3 refer to the DC step discussed earlier in this section. Line 4 returns an α-optimal solution defined in the postcondition of the algorithm.

Algorithm DC(C, n, S, f_C, α)
Input: $C, \mathscr{S}, \alpha$.
Output: $S^\alpha \in \mathscr{S}$ such that $f_C(S^\alpha) \leq \min\{f_C(S) \mid S \in \mathscr{S}\} + \alpha$.
Code:
```
 1 begin
 2        find a feasible instance C_F^α to approximation
                  problem (1.7);
 3        if ρ(C, C_F^α) ≤ α
 4             return S^α;
 5        else
 6        begin
 7             partition 𝒮 into subsets 𝒮_i for i = 1,...,k;
 8             S_i^α := DC(C, n, 𝒮_i, f_C, α) for i = 1,...,k;
 9             return the best solution among S_i^α for i = 1,...,k;
10        end;
11 end.
```

Fig. 1.5 A data-correcting algorithm for combinatorial optimization problems with a minimization objective

Recall that, in the ATSP example, the set $\mathscr{R}$ of polynomial instances, used in the DC algorithm, is the set of distance matrices for which the Hungarian Algorithm returns a tour (cyclic permutation). As a proximity measure we use ρ_2 from formula (1.6). The branching rule makes use of an arc e, yielding a maximum contribution c_e towards the value of the proximity measure, i.e.,

$$c_e \in \arg\max\left\{|c_{ij} - c_{ij}^F| : i, j = 1, \ldots, m\right\}, \tag{1.8}$$

with $[c_{ij}]$ the current input instance and $[c_{ij}^F]$ its "polynomially solvable coworker." Let $K(h)$ and $K(t)$ be subtours from the AP optimal solution that contains either the head h or the tail t of an arc $e = (h, t)$ used for patching $K(h)$ and $K(t)$ (see Fig. 1.4). The current set of Hamiltonian tours $\mathscr{S}$ is partitioned into subsets $\mathscr{S}(e) = \{S \in \mathscr{S} \mid e \notin S\}$ for all arcs e from the shortest (in terms of the number of arcs) subtour among $K(h)$ and $K(t)$. A motivation why branching by excluding arcs is better than branching by including arcs is presented in [51] and [70].

For example (see Fig. 1.4), the maximum contribution c_{24} towards the value of the proximity measure $\rho(C, C_F)$ is found on the arc $(2, 4)$. Since the lengths of both subtours $K(2) = (1231)$ and $K(4) = (4564)$ are the same, we may branch by deleting all arcs either from $K(2)$ or from $K(4)$.

The current matrix C is adapted accordingly. Namely, including an arc $e = (i, j)$ in any feasible solution is equivalent to deleting the row i and column j from C, and excluding an arc $e = (i, j)$ from any feasible solution is equivalent to setting c_{ij} to infinity.

We illustrate the execution of the DC algorithm in case of the 8-city instance from Table 1.1 with accuracy parameter $\alpha = 0$ (see [71]). This instance is considered in [6].

Table 1.1 8-city ATSP instance

		1	2	3	4	5	6	7	8
	1	–	2	11	10	8	7	6	5
	2	6	–	1	8	8	4	6	7
	3	5	12	–	11	8	12	3	11
$C=$	4	11	9	10	–	1	9	8	10
	5	11	11	9	4	–	2	10	9
	6	12	8	5	2	11	–	11	9
	7	10	11	12	10	9	12	–	3
	8	7	10	10	10	6	3	1	–

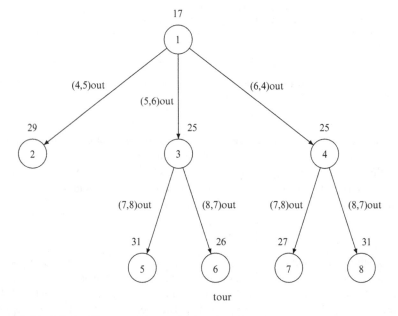

Fig. 1.6 The DC algorithm tree with non-disjoint subproblems

The original problem is the problem at the root node of the solution tree in Fig. 1.6. We call this: subproblem 1. The nodes are labeled according to the order in which the problems are evaluated. The Hungarian Algorithm starts with subtracting its smallest entry from each row (see, e.g., [112]). Similarly, for the columns. We obtain the matrix C^1 in Table 1.2. Its optimal AP solution is given by the entries that are boxed in, namely $S_R^1 = (1231)(4564)(787)$. The subtours are patched and result in the tour $S_F^1 = (123786451)$ with $f_C(S_F^1) = 26$. The superindex of S_F^i corresponds to the label of the subproblem. The cost of patching these subtours is 9, so that the cost of the patched tour exceeds the cost of the AP optimal solution by 9 (see Table 1.4). We next construct the corrected matrix that has S_F^1 as its optimal solution. This is done by decreasing the cost of the arc $(5,1)$ by 7 units and the cost

Table 1.2 ATSP instance after subtracting the row and column minima

		1	2	3	4	5	6	7	8
	1	–	0	9	8	6	5	4	3
	2	3	–	0	7	7	3	5	6
	3	0	9	–	8	5	9	0	8
$C^1 =$	4	8	8	9	–	0	8	7	9
	5	7	9	7	2	–	0	8	7
	6	8	6	3	0	9	–	9	7
	7	5	8	9	7	6	9	–	0
	8	4	9	9	9	5	2	0	–

Table 1.3 The corrected matrix at the root node

		1	2	3	4	5	6	7	8
	1	–	2	11	10	8	7	6	5
	2	6	–	1	8	8	4	6	7
	3	5	12	–	11	8	12	3	11
$C_1^F =$	4	11	9	10	–	1	9	8	10
	5	4	11	9	4	–	2	10	9
	6	12	8	5	2	11	–	11	9
	7	10	11	12	10	9	12	–	3
	8	7	10	10	10	6	1	1	–

Table 1.4 The DC solutions for the 8-city ATSP

i	S_R^i	$f_C(S_R^i)$	S_F^i	$f_C(S_F^i)$	$\rho_2(C,C^F)$
1	(1231)(4564)(787)	17	(123786451)	26	9
2	(1231)(478564)	29	Fathomed		
3	(12631)(454)(787)	25	(126453781)	31	6
4	(12631)(454)(787)	25	(126453781)	31	6
5	(12371)(45864)	31	Fathomed		
6	(123786451)	26	Fathomed		0
7	(1231)(456874)	31	Fathomed		
8	(121)(37863)(454)	27	Fathomed		

of the arc $(8,6)$ by 2 units, and leaving all other arc costs unchanged. This leads to the corrected matrix C_1^F (see Table 1.3).

Clearly, $\rho_2(C,C_1^F) = 7 + 2 = 9$, which is more than the prescribed accuracy $\alpha = 0$. Thus we need to branch.

The largest contribution towards the value of $\rho_2(C,C_1^F)$ is the correction of the cost of arc $(5,1)$. So we branch on all arcs from subtour $(4,564)$ and obtain three subproblems, namely $\mathscr{S}_2 = \mathscr{S}(4,5)$, $\mathscr{S}_3 = \mathscr{S}(5,6)$, and $\mathscr{S}_4 = \mathscr{S}(6,4)$. (Note that subtour $(4,564)$ corresponds to the head of arc $(5,1)$, and subtour $(1,231)$ to the tail of arc $(5,1)$. Since the lengths of both subtours $(1,231)$ and $(4,564)$ are the same, an alternative choice is subtour $(1,231)$.) It is clear that $\mathscr{S}_2 \cup \mathscr{S}_3 \cup \mathscr{S}_4 = \mathscr{S}_1$, although not each pair of subsets has nonempty intersection. A nonoverlapping representation of a set of subproblems can be found in [108].

Now we solve the AP for each of the three subproblems, and find that $f_C(S_R^2) = 29$, $f_C(S_R^3) = 25$, and $f_C(S_R^4) = 25$. Subproblem 2 can be fathomed, since $f_C(S_R^2) = 29 > f_C(S_F^1) = 26$. The three subtours of the solution $S_R^3 = (12631)(454)(787)$ are patched, yielding $S_F^3 = (126453781)$ with $f_C(S_F^3) = 31$, and the cost of patching decreased from 9 to 6. For subproblem 4 the same solutions $S_R^4 = (12631)(454)(787)$, $S_F^4 = (126453781)$ are found, with $f_C(s_F^4) = 31$ and patching cost 6. Now the largest contribution towards the value of $\rho_2(C, C_3^F)$ is the cost correction of the arc $(8, 1)$. The DC algorithm branches on the arcs of subtour (787), giving rise to subproblems $\mathscr{S}_5 = \mathscr{S}(7, 8)$ and $\mathscr{S}_6 = \mathscr{S}(8, 7)$. The AP solution of subproblems 5 and 6 are $f_C(S_R^5) = 31$ and $f_C(S_R^6) = 26$, respectively. Both are fathomed. Since $S_R^3 = S_R^4$, the DC algorithm creates for subproblem 4 two subproblems 7 and 8 with $f_C(S_R^7) = 31$ and $f_C(S_R^8) = 27$, respectively. So both are fathomed. Therefore, the DC algorithm has found the optimal solution $f_C(S^\alpha) = f_C(S_F^1) = 26$ with $\alpha = 0$. Note that 8 subproblems are solved. The results are listed in Table 1.4; the corresponding tree of solutions is shown in Fig. 1.6.

Note that if $\alpha = 1$, then we could stop the solution process after solving the first four problems, since the value of the current lower bound was equal to $f_C(S_R^3) = f_C(S_R^4) = 25$.

As we have seen the optimal solutions $S_R^3 = S_R^4 = (12631)(454)(787) \in \mathscr{S}_3 \cap \mathscr{S}_4 \neq \emptyset$ are the same for subproblems 3 and 4 because their sets of feasible solutions have a nonempty intersection. A set of mutually disjoint solutions (subproblems) (see, e.g., [136]) can be constructed as follows (see Fig. 1.7): $\mathscr{A}_2 = \{S : (5,6) \notin S\}$, $\mathscr{A}_3 = \{S : (5,6) \in S \text{ and } (4,5) \notin S\}$, and $\mathscr{A}_4 = \{S : (5,6), (4,5) \in S \text{ and } (6,4) \notin S\}$ with the corresponding optimal solutions $f_C(S_R^2) = f_C[(12631)(454)(787)] = 25$, $f_C(S_R^3) = f_C[(1231)(478564)] = 29$, $f_C(S_R^4) = f_C[(1245631)(787)] = 27$, the same patched solution $f_C(S_F^1) = 26$, and the same corrected matrix C_1^F. Hence, subproblems 3 and 4 are fathomed. If $\alpha = 1$, then we could stop the solution process after solving the first four subproblems, otherwise (if $\alpha = 0$) we patch the solution S_R^2 into $f_C(S_F^2) = f_C(127964531) = 31$ and split subproblem 2 into two subproblems 5 and 6 with the corresponding $\mathscr{A}_5 = \{S : (5,6), (7,8) \notin S\}$, $\mathscr{A}_6 = \{S : (7,8) \in S \text{ and } (5,6), (8,7) \notin S\}$, and $f_C(S_R^5) = f_C[(12371)(45864)] = 31$, $f_C(S_R^6) = f_C[(123786451)] = 26$. Therefore, the DC algorithm has found the optimal solution by solving 6 subproblems. Note that in both implementations (with and without overlapping sets of feasible solutions) the DC algorithm has the largest contribution 4 towards the values of $\rho(C, C_3^F)$ and $\rho(C, C_2^F)$ for the corresponding inserted arcs $(8, 1)$ and $(2, 8)$, respectively, and indicates for branching the same shortest subtour (787).

The previous example shows that the DC algorithm can be an attractive alternative to usual BnB algorithms (see, e.g., [6]). In the next section we report computational experiences with the data correcting algorithm on ATSP instances.

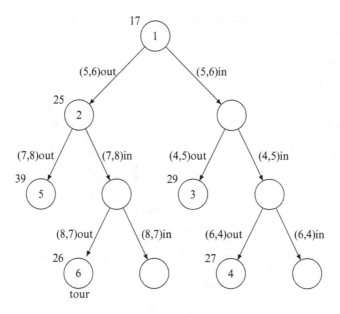

Fig. 1.7 The DC algorithm tree with disjoint subproblems

1.4 Preliminary Computational Experience with ATSP Instances

In this section we demonstrate the effectiveness of the DC algorithm on some benchmark ATSP instances from TSPLIB [121]. TSPLIB has 27 ATSP instances, out of which we have chosen 12 which could be solved to optimality within 5 h using a basic BnB algorithm. Eight of these belong to the "*ftv*" class of instances, while four belong to the "*rbg*" class. We implemented the DC algorithm in C and ran it on an Intel Pentium computer with 666 MHz and 128 MB RAM.

The results of our experiments are presented graphically in Figs. 1.8–1.11. In computing accuracies (Figs. 1.8 and 1.10), we have plotted the accuracy and deviation of the solution output by the data correcting algorithm from the optimal (called "achieved accuracy" in the figures) as a fraction of the cost of an optimal solution to the instance. We observed that for each of the 12 instances that we studied, the achieved accuracy is consistently less than 80% of the prespecified accuracy.

There was a wide variation in the CPU time required to solve the different instances. For example, $ftv70$ required 17206 s to solve to optimality, while $rbg323$ required just 5 s. Thus, in order to maintain uniformity while demonstrating the variation in execution times with respect to changes in α values, we represented the execution times for each instance for each α value as a percentage of the execution time required to solve that instance to optimality. Notice that for all the ftv instances

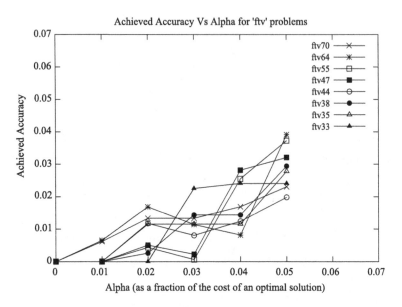

Fig. 1.8 Accuracy achieved versus α for *ftv* instances

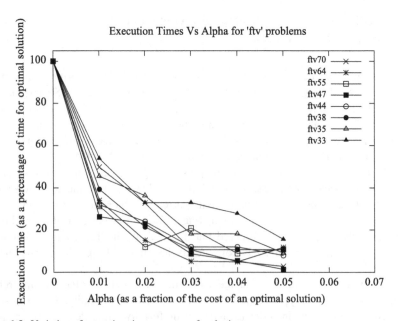

Fig. 1.9 Variation of execution times versus α for *ftv* instances

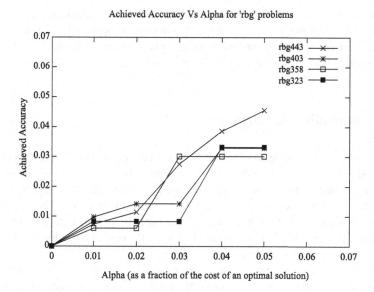

Fig. 1.10 Accuracy achieved versus α for *rbg* instances

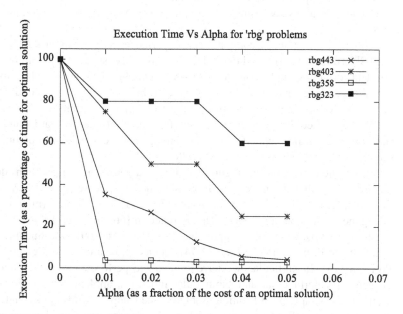

Fig. 1.11 Variation of execution times versus α for *rbg* instances

when α was 5 % of the cost of the optimal solution, the execution time was reduced to 20 % of that required to solve the respective instance to optimality. The reductions in execution times for *rbg* instances were equally steep, with the exception of *rbg*323 which was anyway an easy instance to solve.

1.5   Concluding Remarks

In this chapter we provide an introduction to the concept of Data Correcting (DC), a method in which our knowledge of polynomially solvable special cases in a given problem domain is utilized to obtain near-optimal solutions with the prespecified performance guarantees within relatively short execution times. The algorithm makes use of the fact that even if the cost of an optimal solution to a given instance is not known, it is possible to compute a bound on the cost of the solution based on the cost of an optimal solution to another instance.

In Sect. 1.1, we describe the DC process on a single variable real-valued function. Most of the terminology used in data correcting is defined in this section. We also provide a pseudocode for a DC algorithm for a general real-valued function and an example demonstrating the algorithm. In Sect. 1.2, we show how the DC approach can be used to solve NP-hard combinatorial optimization problems. It turns out that it fits nicely into the framework of BnB. We also provide a pseudocode for an algorithm applying DC, on a combinatorial optimization problem with min-sum objective, and show, using an example, how the algorithm works on the ATSP in Sect. 1.3.

In Sect. 1.4 we describe computational experiences with benchmark ATSPs from the TSPLIB (see [121]). We show that the deviation in cost of the solutions output by our data correcting implementation from the optimal is about 80 % of the allowable deviation, and the time required to solve the problem at hand to 95 % optimality is about 20 % of the time required to solve the problem to optimality.

We have used DC primarily for solving some of the NP-hard combinatorial optimization problems. Our choice among many examples of combinatorial optimization problems is motivated as follows:

One of the chosen problems should represent a wide range of problems defined on the set of all permutations and another one—on the set of all subsets of a finite set, i.e., we have chosen the ATSP [54] and maximization (minimization) of a general submodular (supermodular) function [66] specified by means of the quadratic cost partition [72], Simple Plant Location [1, 60, 69], assortment problem [56, 57, 61–63], *p*-Median Problem [2, 74], and its application to the Cell Formation Problem [75]. In particular, we have studied the performance of this algorithm on general supermodular and submodular functions, applied it to quadratic cost partition and simple plant location problems (see Chap. 3 of this book), and in this chapter, on the ATSP. Much research remains to be done on testing the performance of this approach on other hard combinatorial problems.

Chapter 2
Maximization of Submodular Functions: Theory and Algorithms

2.1 Introduction

In this chapter we give some theoretical results fundamental to the problem of finding a global maximum of a general submodular (or, equivalently, global minimum of a general supermodular) set function (see [73] which we call the problem of maximization of submodular functions (PMSFs) (following [102]). By a set function we mean a mapping from 2^N to the real numbers, where $N = \{1, 2, \ldots, n\}$. Another well-known term for an arbitrary set function is a *pseudo-Boolean function* (see [84], [20, 21]) which is a mapping from $\{0, 1\}^n$ to the real numbers. PMSF is known to be NP-hard, though the corresponding minimization problem is known to be polynomially solvable (see, e.g., [127]). Enormous interest in studying PMSF arises from the fact that several classes of important combinatorial optimization problems belong to PMSF, including the Simple or "Uncapacitated" Plant (Facility) Location Problem (SPLP) and its competitive version (see [11]), the quadratic cost partition (QCP) problem with nonnegative edge weights, and its special case—the Max-Cut Problem, the generalized transportation problem [109, 110]. Many models in mathematics [103], including the rank function of elementary linear algebra, which is a special case of matroid rank functions (see [42, 45]), require the solution of a PMSF.

Although the general problem of the maximization of a submodular function is known to be NP-hard, there has been a sustained research effort aimed at developing practical procedures for solving medium and large-scale instances of the PMSF. Often the approach taken has been problem specific, and supermodularity of the underlying objective function has been only implicit to the analysis. For example, Barahona et al. [7] have addressed the max-cut problem from the point of view of polyhedral combinatorics and developed a branch and cut algorithm, suitable for applications in statistical physics and circuit layout design. Beasley [9] applies Lagrangean heuristics to several classes of location problems including SPLPs and

B. Goldengorin and P.M. Pardalos, *Data Correcting Approaches in Combinatorial Optimization*, SpringerBriefs in Optimization, DOI 10.1007/978-1-4614-5286-7_2, © Boris Goldengorin, Panos M. Pardalos 2012

reports results of extensive experiments on a Cray supercomputer. Lee et al. [102] have made a study of the QCP problem of which max-cut with nonnegative edge weights is a special case, again from the standpoint of polyhedral combinatorics.

There have been fewer published attempts to develop algorithms for minimization of a general supermodular function. We believe that the earliest attempt to exploit supermodularity is the work of [118], who identified a supermodular structure in their study of railway timetabling. Their procedure was subsequently published by Cherenin [29] as the "method of successive calculations." Their algorithm however is not widely known in the West [4] where, as far as we are aware of, the only general procedures that have been studied in depth are the greedy approximation algorithm from [109], and the algorithm for maximization of submodular functions subject to linear constraints from [110]. In a comment to a note by Frieze [46], Babayev [4] demonstrated that Frieze's two rules: OP1 and OP2, developed to accelerate a BnB algorithm for the SPLP were a consequence of Cherenin's theorem for PMSF [29]. Note that Alcouffe and Muratet's [3] algorithm is based on a special case of Cherenin's [29] "method of successive calculations."

Indeed the only practical algorithmic implementation known in the West appears to be the "accelerated greedy" (AG) algorithm of [105], which has been applied to optimal planning and design of telecommunication networks. We note that the AG algorithm has also been applied to the problem of D-optimal experimental design [125]; see also Ko et al. [92] and Lee [101] for further examples of "hard" D-optimal design problems in environmental monitoring. In [50] an optimal algorithm is constructed with exponential time complexity for the well-known Shannon max–min problem. This algorithm is applied to the maximization of submodular functions subject to a convex set of feasible solutions, and to the problem of—what is known as—decoding monotonic Boolean functions.

In this chapter we present an elegant key theorem of Cherenin, which provides the basis of excluding rules, and in particular, for the justification of the Preliminary Preservation (Dichotomy) algorithm. We generalize Cherenin's excluding rules in the form of "preservations rules" which will be used in Chap. 3. Moreover, our preservation rules can be used for implicit enumeration of subproblems in a BnB approach for solving PMSF.

The chapter is organized as follows. In Sect. 2.2 we motivate a theoretical development of these rules by presenting some important results on the structure of local and global maxima for submodular functions by Cherenin [29] and Khachaturov [89, 90]. In this section a fundamental theorem of Cherenin is stated, which provides the basis of "the method of successive calculations." Section 2.2 also contains an important characterization of local maxima as disjoint components of "strict" and "saddle" vertices which greatly assists the understanding of the difference between the properties of Cherenin's "excluding rules" and our "preservation rules" discussed in Sect. 2.3. In Sect. 2.4 we present our main Theorem 2.8 from which generalized bounds for implicit enumeration can be derived, and allow the rules of Sect. 2.3 to be extended to other cases (ε-optimality). We present the two different representations (a) and (b) of the partition of the current set of feasible solutions (vertices) defined by a strictly inner vertex with respect to this set.

By using our main Theorem 2.8 and representations (a) and (b), we prove the correctness of Cherenin's excluding rules in the form of our preservation rules. These rules are the basis of Cherenin's preliminary preservation algorithm (PPA) [118]. We introduce the so-called *nonbinary branching rules*, based on Theorem 2.8 in Sect. 2.6. Nonbinary branching rules are illustrated by an instance of the SPLP. In Sect. 2.5 we outline the main steps of the PPA and illustrate how our new preservation rules (see Corollary 2.6) can be applied to a small example of the SPLP. We show that if the PPA terminates with a global maximum, then the given submodular function has exactly one strict component. Section 2.7 gives a number of concluding remarks.

2.2   The Structure of Local and Global Maxima of Submodular Set Functions

In this section we present results of Cherenin–Khachaturov (see [29, 89]) which are hardly known in the Western literature (see also [4]).

Let z be a real-valued function defined on the power set 2^N of $N = \{1, 2, \ldots, n\}$; $n \geq 1$. For each $S, T \in 2^N$ with $S \subseteq T$, define

$$[S, T] = \{I \in 2^N \mid S \subseteq I \subseteq T\}.$$

Note that $[\emptyset, N] = 2^N$. Any *interval* $[S, T]$ is, in fact, a *subinterval* of $[\emptyset, N]$ if $\emptyset \subseteq S \subseteq T \subseteq N$; notation $[S, T] \subseteq [\emptyset, N]$. In this book we mean by an interval always a subinterval of $[\emptyset, N]$. Throughout this book we consider a set of PMSFs defined on any interval $[S, T] \subseteq [\emptyset, N]$ as follows:

$$\max\{z(I) \mid I \in [S, T]\} = z^*[S, T], \text{ for all } [S, T] \subseteq [\emptyset, N].$$

The function z is called *submodular* on $[S, T]$ if for each $I, J \in [S, T]$ it holds that

$$z(I) + z(J) \geq z(I \cup J) + z(I \cap J).$$

Expressions of the form $S \setminus \{k\}$ and $S \cup \{k\}$ will be abbreviated to $S - k$ and $S + k$.

The following theorem presented in [109] gives a number of equivalent formulations for submodular functions which is useful for a clearer understanding of the concept of submodularity. Since sometime we use the incremental or decremental value of $z(S)$, we define $d_j^+(S) = z(S + j) - z(S)$ and $d_j^-(S) = z(S - j) - z(S)$.

Theorem 2.1. *All the following statements are equivalent and define a submodular function.*

(i) $z(A) + z(B) \geq z(A \cup B) + z(A \cap B), \ \forall A, B \subseteq N.$

(ii) $d_j^+(S) \geq d_j^+(T), \ \forall S \subseteq T \subseteq N \text{ and } j \in N \setminus T.$

(iii) $d_j^+(S) \geq d_j^+(S+k)$, $\forall S \subseteq N$ and $j \in N \setminus (S+k)$

$$\text{and } k \in N \setminus S.$$

(iv) $z(T) \leq z(S) + \sum_{j \in T \setminus S} d_j^+(S)$, $\quad \forall S \subseteq T \subseteq N$.

(v) $z(S) \leq z(T) + \sum_{j \in T \setminus S} d_j^-(T)$, $\quad \forall S \subseteq T \subseteq N$.

As an example consider the QCP problem, for which it is well known that the objective function $z(Q)$ is a submodular function (see e.g., [102]). For given real numbers p_i and nonnegative real numbers q_{ij} with $i, j \in N$, the QCP is the problem of finding a subset Q of N such that the weight $z(Q) = \sum_{i \in Q} p_i - \frac{1}{2} \sum_{i, j \in Q} q_{ij}$ is as large as possible. Let N be the vertex set, $E \subseteq N \times N$ the edge set of an edge-weighted graph $G = (N, E)$, and $w_{ij} \geq 0$ are edge weights. For each $Q \subseteq N$, the cut $\delta(Q)$ is defined as the edge set for which each edge has one end in Q and the other one in $N \setminus Q$. It is easy to see that the Max-Cut Problem with nonnegative edge weights is a QCP where $p_i = \sum_{j \in N} w_{ij}$ and $q_{ij} = 2w_{ij}$, for $i, j \in N$.

Lemma 2.1. *The objective $z(S)$ of the QCP problem is submodular.*

Proof. According to Theorem 2.1(iii) a function is submodular if

$$d_l^+(S) \geq d_l^+(S+k), \ \forall S \subseteq N \text{ and } l \in N \setminus (S+k) \text{ and } k \in N \setminus S.$$

Substituting $d_l^+(S) = z(S+l) - z(S)$ we get

$$z(S+l) - z(S) \geq z(S+k+l) - z(S+k)$$

Substituting $z(S) = \sum_{i \in S} p_i - \frac{1}{2} \sum_{i, j \in S} q_{ij}$ gives

$$\sum_{i \in S+l} p_i - \frac{1}{2} \sum_{i, j \in S+l} q_{ij} - \left(\sum_{i \in S} p_i - \frac{1}{2} \sum_{i, j \in S} q_{ij} \right)$$

$$\geq \sum_{i \in S+k+l} p_i - \frac{1}{2} \sum_{i, j \in S+k+l} q_{ij} - \left(\sum_{i \in S+k} p_i - \frac{1}{2} \sum_{i, j \in S+k} q_{ij} \right)$$

Canceling out terms involving p_i, we obtain

$$- \sum_{i, j \in S+l} q_{ij} + \sum_{i, j \in S} q_{ij} \geq - \sum_{i, j \in S+k+l} q_{ij} + \sum_{i, j \in S+k} q_{ij}$$

This result, after some bookkeeping, implies

$$q_{kl} + q_{lk} \geq 0$$

Since q_{ij} is nonnegative for all $i, j \in N$, the proof is completed.   □

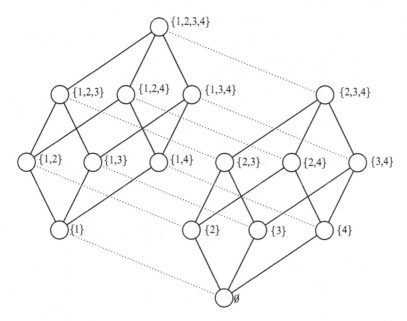

Fig. 2.1 The Hasse diagram of $\{1,2,3,4\}$

Hence, the QCP problem is a special case of the problem of maximizing a submodular function.

A subset $L \in [\emptyset, N]$ is called a *local maximum* of z if for each $i \in N$,

$$z(L) \geq \max\{z(L-i), z(L+i)\}.$$

A subset $S \in [\emptyset, N]$ is called a *global maximum* of z if $z(S) \geq z(I)$ for each $I \in [\emptyset, N]$. We will use the Hasse diagram (see e.g., [76] and Fig. 2.1) as the ground graph $G = (V, E)$ in which $V = [\emptyset, N]$ and a pair (I, J) is an edge if either $I \subset J$ or $J \subset I$, and $|I \setminus J| + |J \setminus I| = 1$.

The graph $G = (V, E)$ is called *z-weighted* if the weight of each vertex $I \in V$ is equal to $z(I)$; notation $G = (V, E, z)$. In terms of $G = (V, E, z)$ the PMSF means finding a vertex $S \in V$ of the weight $z(S)$ which is as large as possible. An example of the weighted G with $N = \{1, 2, 3, 4\}$ is depicted in Fig. 2.2, where the weight $z(I)$ is indicated inside the corresponding vertex I.

Here among others the vertices $\{1, 2, 3\}$ and $\{4\}$ are local maxima, and $\{4\}$ is a global maximum (see Fig. 2.2).

A sequence $\Gamma = (I^0, I^1, \ldots, I^n)$ of subsets $I^t \in 2^N$, $t = 0, 1, \ldots, n$ such that $|I^t| = t$ and

$$\emptyset = I^0 \subset I^1 \subset I^2 \subset \ldots \subset I^t \subset \ldots \subset I^{n-1} \subset I^n = N$$

is called a *chain* in $[\emptyset, N]$. An example of the chain $\emptyset \subset \{2\} \subset \{2, 4\} \subset \{1, 2, 4\} \subset \{1, 2, 3, 4\}$ in Fig. 2.3 is shown.

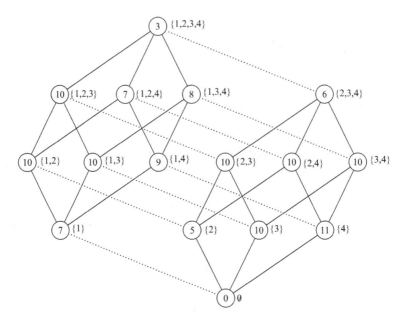

Fig. 2.2 Example of local maxima $\{1,2\}$, $\{1,2,3\}$, $\{1,3\}$, $\{2,3\}$, $\{3\}$, and the global maximum $\{4\}$ on the Hasse diagram

Similarly, a chain of any interval $[S,T]$ can be defined. A submodular function z is *nondecreasing (nonincreasing)* on the chain Γ if $z(I^l) \leq z(I^m)$ ($z(I^l) \geq z(I^m)$) for all l, m such that $0 \leq l \leq m \leq n$; concepts of *increasing, decreasing, and constant* (signs, respectively, $<,>,=$) are defined in an obvious manner (see, for example, Fig. 2.4).

The following theorem (see [29]) shows the quasiconcavity of a submodular function on any chain that includes a local maximum (see Fig. 2.5).

Theorem 2.2. *Let z be a submodular function on 2^N and let L be a local maximum. Then z is nondecreasing on any chain in $[\emptyset,L]$, and nonincreasing on any chain in $[L,N]$.*

Proof. We show that z is nondecreasing on any chain in $[\emptyset,L]$. If either $L = \emptyset$ (we obtain the nonincreasing case) or $|L| = 1$, the assertion is true, since L is a local maximum of z. So, let $|L| > 1$ and $I,J \in [\emptyset,L]$ such that $J = I + k, k \in L \setminus I$.

Note that $\emptyset \subseteq \ldots \subseteq I \subset J \subseteq \ldots \subset L$. The submodularity of z implies $z(J)+z(L-k) \geq z(I)+z(L)$, or $z(J) - z(I) \geq z(L) - z(L-k)$. Since L is a local maximum, $z(L) - z(L-k) \geq 0$. Hence $z(J) \geq z(I)$, and we have finished the proof of the nondecreasing case. The proof for $[L,N]$ is similar. $\square$

Corollary 2.1. *Let z be a submodular function on 2^N and let L_1 and L_2 be local maxima with $L_1 \subseteq L_2$. Then z is a constant on $[L_1,L_2]$, and every $L \in [L_1,L_2]$ is a local maximum of z.*

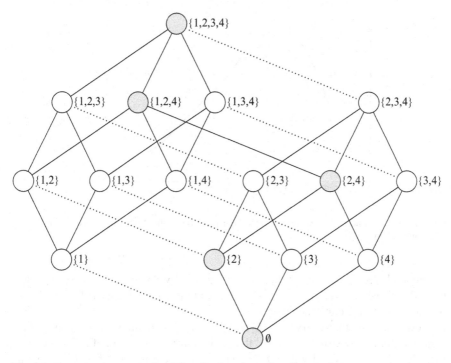

Fig. 2.3 Example of the chain $\emptyset \subset \{2\} \subset \{2,4\} \subset \{1,2,4\} \subset \{1,2,3,4\}$ in the Hasse diagram of $\{1,2,3,4\}$

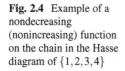**Fig. 2.4** Example of a nondecreasing (nonincreasing) function on the chain in the Hasse diagram of $\{1,2,3,4\}$

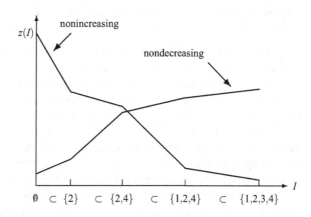

Proof. First we show that z is a constant function on $[L_1, L_2]$. Let us apply Theorem 2.2 to a chain including $\emptyset \subseteq \cdots \subseteq L_1 \subseteq L_2 \subseteq \cdots \subseteq N$, first to the single (isolated) local maximum L_2 and second to the single local maximum L_1. For the first case we obtain $z(\emptyset) \leq \cdots \leq z(L_1) \leq \cdots \leq z(I) \leq z(L_2)$. For any subchain of

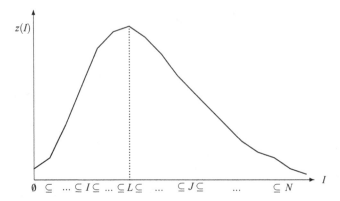

Fig. 2.5 A quasiconcave behavior of a submodular function on the chain with a local maximum L (Cherenin's theorem)

the interval $[L_1, L_2]$ we have $z(L_1) \leq \cdots \leq z(L_2)$. By the same reasoning for the second case we have $z(L_1) \geq \cdots \geq z(L_2)$. Combining both sequences of inequalities we have proved the constancy of z.

Now we show that every $L \in [L_1, L_2]$ is a local maximum of z. Assume the contrary that there exists $L \in [L_1, L_2]$ that is not a local maximum of z. Then either there is a $L - i \notin [L_1, L_2]$ with $z(L) < z(L - i)$ or there is a $L + i \notin [L_1, L_2]$ with $z(L) < z(L + i)$. In the first case we get accordingly the definition of submodularity $z(L) + z(L_2 - i) \geq z(L - i) + z(L_2)$ or $z(L) - z(L - i) \geq z(L_2) - z(L_2 - i) \geq 0$. This contradicts $z(L) < z(L - i)$. In the second case a similar argument holds by using L_1 instead of L_2. □

In Corollary 2.1 we have indicated two important structural properties of a submodular function considered on intervals whose end points are local maxima. Namely, on such an interval a submodular function preserves a constant value and every point of this interval is a local maximum. It will be natural to consider the widest intervals with the above-mentioned properties.

A local maximum $\underline{L} \in 2^N$ ($\overline{L} \in 2^N$) is called a *lower* (respectively, *upper*) *maximum* if there is no other local maximum L such that $L \subset \underline{L}$ (respectively, $\overline{L} \subset L$). For example, in Fig. 2.6 the vertex $\{1, 2, 3\}$ is an upper local maximum and the vertices $\{1, 2\}$, $\{3\}$ are lower local maxima. Note that the vertex $\{3, 4\}$ is not a local maximum. If an interval $[\underline{L}, \overline{L}]$ with $\underline{L} \subseteq \overline{L}$ has as its end points lower and upper maxima, then it is the widest interval on which the submodular function is a constant and each point is a local maximum. We call a pair of intervals $[\underline{L}_i, \overline{L}_i]$ with $\underline{L}_i \subseteq \overline{L}_i$, $i = 1, 2$ *connected* if $[\underline{L}_1, \overline{L}_1] \cap [\underline{L}_2, \overline{L}_2] \neq \emptyset$. The intervals of local maxima form a set of *components of local maxima*. Two intervals belong to the same component if they are connected. Hence, two local maxima L_1 and L_2 are in the same component if there is a path in $G = (V, E, z)$ with end vertices L_1 and L_2, and all intermediate vertices of this path are local maxima.

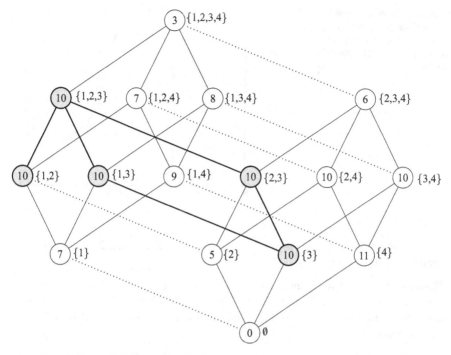

Fig. 2.6 Lower local maxima: $\{1,2\}$, $\{3\}$; upper local maximum: $\{1,2,3\}$; SDC (*shadowed*); global maximum: $\{4\}$

By the following definitions [89] (see also [64]) introduced two kinds of components of subgraphs of local maxima.

Let V_0 be the subset of V corresponding to all local maxima of z and let $H_0 = (V_0, E_0, z)$ be the subgraph of G induced by V_0. This subgraph consists of at least one component. We denote the components by $H_0^j = (V_0^j, E_0^j, z)$, with $j \in J_0 = \{1, \ldots, r\}$. Note that if L_1 and L_2 are vertices in the same component, then $z(L_1) = z(L_2)$.

A component H_0^j is called a *strict local maximum component* (STC) if for each $I \notin V_0^j$, for which there is an edge (I, L) with $L \in V_0^j$, we have $z(I) < z(L)$. A component H_0^j is called a *saddle local maximum component* (SDC) if for some $I \notin V_0^j$, there exists an edge (I, L) with $L \in V_0^j$ such that $z(I) = z(L)$. An example of the SDC defined by two intervals $[\{1,2\}, \{1,2,3\}]$ and $[\{3\}, \{1,2,3\}]$ is shown in Fig. 2.6. The values of a submodular function in Fig. 2.6 are printed inside the vertices. Here a trivial STC by the vertex $\{4\}$ is defined. Note that $\{3,4\}$ is not a local maximum because its neighbor $\{4\}$ is the global maximum with value $z(\{4\}) = 11$.

All vertices in a component H_0^j are local maxima of the same kind. Therefore, the index set J_0 of these components can be split into two subsets: J_1 being the index set of the STCs, and J_2 being the index set of the SDCs.

Fig. 2.7 The behavior of a submodular function on a chain with lower and upper local maxima (Khachaturov's theorem)

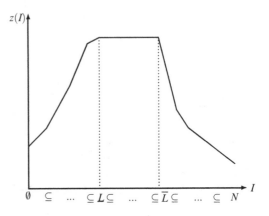

The following theorem of [89] is an application of Theorem 2.2 to the case of a nontrivial STC (see Fig. 2.7).

Theorem 2.3. *Let z be a submodular function on 2^N and let $\underline{L}$ and $\overline{L}$ be lower and upper maxima with $\underline{L} \subseteq \overline{L}$, both located in an STC. Then z is strictly increasing on each subchain $\emptyset \subseteq \cdots \subseteq \underline{L}$ of $[\emptyset, \underline{L}]$, constant on $[\underline{L}, \overline{L}]$, and strictly decreasing on each subchain $\overline{L} \subseteq \cdots \subseteq N$ of $[\overline{L}, N]$.*

Proof. We first show that z is strictly increasing on $[\emptyset, \underline{L}]$. The proof of the strictly decreasing case is similar. If either $\underline{L} = \emptyset$ (we obtain the decreasing case) or $|\underline{L}| = 1$, the assertion is true, since $\underline{L}$ is a local maximum of z. So, let $|\underline{L}| > 1$ and $I, J \in [\emptyset, \underline{L}]$ such that $J = I + k$, $k \in L \setminus I$. Note that $\emptyset \subseteq I \subset J \subseteq \cdots \subseteq \underline{L}$. The submodularity of z implies $z(J) + z(\underline{L} - k) \geq z(I) + z(\underline{L})$, or $z(J) - z(I) \geq z(\underline{L}) - z(\underline{L} - k)$. Since $\underline{L} \in V_0^j$ for some $j \in J_1$, $z(\underline{L}) - z(\underline{L} - k) > 0$. Hence $z(J) > z(I)$, and we have finished the proof of the strictly increasing case.

The property that z is constant on $[\underline{L}, \overline{L}]$ follows from Corollary 2.1.   □

Note that $\underline{L}$ and $\overline{L}$ need not be lower and upper maxima in Theorem 2.3. It is clear from the proof of Theorem 2.3 that any pair of embedded local maxima L_1 and L_2 located on a chain $\emptyset \subseteq \cdots \subseteq L_1 - i \subset L_1 \subseteq \cdots \subseteq L_2 \subset L_2 + k \subseteq \cdots \subseteq N$ such that $z(L_1 - i) < z(L_1)$ and $z(L_2 + k) < z(L_2)$ will imply that z is strictly increasing on each subchain $\emptyset \subseteq \cdots \subseteq L_1 - i \subset L_1$ and strictly decreasing on each subchain $L_2 \subset L_2 + k \subseteq \cdots \subseteq N$. We call such a local maximum *boundary local maximum*. In other words, a boundary local maximum is connected with vertices outside the component.

Lemma 2.2. *Let $L \in V_0^j$ for some $j \in J_1$, and let I satisfy $z(I) = z(L)$ with $(I, L) \in E$. Then $I \in V_0^j$ for the same $j \in J_1$.*

Proof. Let $L \in V_0^j$ for some $j \in J_1$. If $I \notin V_0^j$, then $z(I) < z(L)$, since $(I, L) \in E$ and L is a local maximum of the STC.   □

Khachaturov [89] has observed that any global maximum belongs to a STC.

Theorem 2.4. *Let S be a global maximum of the submodular function z defined on 2^N. Then $S \in V_0^j$ for some $j \in J_1$.*

Proof. Suppose, to the contrary, that $S \in V_0^i$ with $i \in J_2$. Then there exists an $I \in V \setminus V_0$, adjacent to some $J \in V_0^i$ with $z(I) = z(J)$. This I is not a local maximum and hence I has an adjacent vertex M with $z(M) > z(I)$. Thus $z(S) = z(J) = z(I) < z(M)$, contradicting the assumption that S is a global maximum of z. □

Theorem 2.4 implies that we may restrict the search for a global maximum of a submodular function z to STCs. Based on Corollary 2.1, and definitions of strict and saddle components, we can represent each component of local maxima as a maximal connected set of intervals whose end points are lower and upper local maxima.

2.3 Excluding Rules: An Old Proof

There are two "excluding rules" (see [3, 46, 118]) that can be used to eliminate certain subsets from $[\emptyset, N]$ when determining a global maximum of a submodular function.

Theorem 2.5. *Let z be a submodular function on $[\emptyset, N]$ and V_0^j with $j \in J_0$ be the components of local maxima. Then the following assertions hold.*

(a) *First Strict Excluding Rule (FSER).*
   If for some T_1 and T_2 with $\emptyset \subseteq T_1 \subset T_2 \subseteq N$ we have $z(T_1) > z(T_2)$, then $V_0^j \cap [T_2, T] = \emptyset$ for all $j \in J_0$.
(b) *Second Strict Excluding Rule (SSER).*
   If for some S_1 and S_2 with $\emptyset \subseteq S_1 \subset S_2 \subseteq N$ we have $z(S_1) < z(S_2)$, then $V_0^j \cap [S, S_1] = \emptyset$ for all $j \in J_0$.

Proof. We prove case (a) because a proof of case (b) is similar. Let us consider a chain $\emptyset \subseteq \cdots \subseteq T_1 \subset T_2 \subseteq L \subseteq T \subset \ldots \subset N$ with $L \in V_0^j \cap [T_2, T] \neq \emptyset$ for some $j \in J_0$. Applying Theorem 2.2 to the subchain $\emptyset \subseteq \cdots \subseteq T_1 \subset T_2 \subseteq L$, we have $z(\emptyset) \leq \cdots \leq z(T_1) \leq z(T_2) \leq z(L)$ which contradicts $z(T_1) > z(T_2)$. □

This theorem states that by applying the strict rules we do not exclude any local maximum. In other words, we preserve all local maxima. In the following theorem of [89] we will see that applying excluding rules with nonstrict inequalities (nonstrict rules) will preserve at least one local maximum of each STC. We will call such a maximum a *representative* of the STC.

Theorem 2.6. *Let z be a submodular function on $[S, T] \subseteq [\emptyset, N]$ and for every $j \in J_1, V_0^j \cap [S, T] \neq \emptyset$. Then the following assertions hold.*

(a) *First Excluding Rule (FER).*
   If for some T_1 and T_2 with $S \subseteq T_1 \subset T_2 \subseteq T$ holds that $z(T_1) \geq z(T_2)$, then $V_0^j \cap ([S, T] \setminus [T_2, T]) \neq \emptyset$ for all $j \in J_1$.

(b) *Second Excluding Rule (SER).*

   If for some S_1 and S_2 with $S \subseteq S_1 \subset S_2 \subseteq T$ holds that $z(S_1) \leq z(S_2)$, then $V_0^j \cap ([S,T] \setminus [S,S_1]) \neq \emptyset$ for all $j \in J_1$.

Proof. We prove case (a) because the proof of case (b) is similar. Let us consider two cases:

Case 1. $z(T_1) > z(T_2)$. Theorem 2.5 implies that $V_0^j \cap [T_2,T] = \emptyset$ for all $j \in J_0 = J_1 \cup J_2$. Since for every $j \in J_1, V_0^j \cap [S,T] \neq \emptyset$ and $[T_2,T] \subset [S,T]$ we have $([S,T] \setminus [T_2,T]) \cap V_0^j \neq \emptyset$ for all $j \in J_1$.

Case 2. $z(T_1) = z(T_2)$. If we can construct a chain through two boundary local maxima L_1 and L_2 that also contains T_1 and T_2, then we have just two possibilities:

1. $L_1 \subseteq T_1 \subset T_2 \subseteq L_2$.
2. All others.

Each case of the possibility (2) contradicts Theorem 2.3. Therefore, $L_1 \subseteq T_1 \subset T_2 \subseteq L_2$, and $L_1 \subseteq T_1 \in ([S,T] \setminus [T_2,T]) \cap V_0^j \neq \emptyset$ for all $j \in J_1$.   □

   In Sect. 2.6 we will give an example of the SPLP in which by application of a nonstrict excluding rule we discard the local minimum $\{2,4\}$ of the corresponding supermodular function. This local minimum is an analogue of the trivial SDC for the corresponding supermodular function.

   By applying Theorem 2.6a (respectively, 2.6b) we can discard $2^{|T \setminus T_2|}$ (respectively, $2^{|S_1 \setminus S|}$) subsets of interval $[T_2,T]$ (respectively, $[S,S_1]$) because this interval does not include a local maximum of any STC from $[S,T]$. If $T_1 = S$ and $T_2 = S+i$ then in case of Theorem 2.6a the interval $[S+i,T]$ can be discarded. If $S_1 = T - i$ and $S_2 = T$ then in case of Theorem 2.6b the interval $[S, T - i]$ can be discarded. These two special cases are important because we may exclude a *half subinterval* of the current interval while we preserve at least one representative from each STC. The rules excluding a half subinterval are called *prime rules*.

   Based on the last special cases of excluding rules, we present Cherenin's Preliminary Preservation (Dichotomy) Algorithm for the maximization of submodular functions in Sect. 2.5. Before we present the dichotomy algorithm we give in Theorem 2.7 an alternative proof of the correctness of these special cases of excluding rules which is based only on Lemma 2.2, the definitions of an STC and a submodular function z. This proof shows that in case of submodular functions the definition of a STC is an insightful notion for understanding the correctness of Cherenin's dichotomy algorithm. Therefore, it is not necessary to use all the statements of the previous section in order to justify both prime rules. In the next section we present a generalization and a simple justification of the same rules.

Theorem 2.7. *Let z be a submodular function on 2^N. Suppose that for $\emptyset \subseteq S \subset T \subseteq N$ and for every $j \in J_1$, $V_0^j \cap [S,T] \neq \emptyset$. Then the following assertions hold.*

(a) *First Prime Excluding Rule (FPER).*

   If for some $i \in T \setminus S$ it holds that $z(S+i) \leq z(S)$, then $[S, T - i] \cap V_0^j \neq \emptyset$ for all $j \in J_1$.

(b) Second Prime Excluding Rule (SPER).
If for some $i \in T \setminus S$ it holds that $z(T - i) \leq z(T)$, then $[S+i,T] \cap V_0^j \neq \emptyset$ for all $j \in J_1$.

Proof. We prove only part (a). The proof of part (b) is similar.

(a) Let $z(S+i) \leq z(S)$ for some $i \in T \setminus S$ and let $G \in V_0^j \cap [S,T]$ for any $j \in J_1$. Then $S \subset G$.

Case 1. $i \in G$. From the definition of submodularity applied to the sets $G - i$ and $S + i$

$$z(G - i) + z(S + i) \geq z(G \cup S + i) + z(S) \Rightarrow$$
$$z(G - i) - z(G \cup S + i) \geq z(S) - z(S + i) \geq 0 \Rightarrow$$
$$z(G - i) \geq z(G \cup S + i) = z(G) \Rightarrow (G \text{ is a local maximum})$$
$$z(G - i) = z(G). \ G \in V_0^j \Rightarrow (\text{by Lemma 2.2})$$
$$G - i \in V_0^j \Rightarrow G - i \in V_0^j \cap [S, T - i] \Rightarrow V_0^j \cap [S, T - i] \neq \emptyset.$$

Case 2. $i \notin G$.
$$i \notin G \Rightarrow G \in V_0^j \cap [S, T - i] \Rightarrow V_0^j \cap [S, T - i] \neq \emptyset. \qquad \square$$

Theorem 2.7a states that if $z(S + i) - z(S) \leq 0$ for some $i \in T \setminus S$, then by preserving the interval $[S, T - i]$ we preserve at least one strict local maximum from each STC, and hence we preserve at least one global maximum from each STC containing a global maximum. Therefore, in this case it is possible to exclude exactly the whole interval $[S+i,T]$ of $[S,T]$ from consideration when searching for a global maximum of the submodular function z on $[S,T] \subseteq [\emptyset,N]$. For example, see Fig. 2.8, if $z(\emptyset) - z(\emptyset + 1) \geq 0$, then the interval $[\{1\},\{1,2,3,4\}]$ can be excluded, i.e., the interval $[\emptyset,\{2,3,4\}]$ will be preserved (FPER). If $z(\{1,2,3,4\}) - z(\{1,2,3,4\} - 1) \geq 0$, then the interval $[\emptyset,\{2,3,4\}]$ can be excluded, i.e., the interval $[\{1\},\{1,2,3,4\}]$ will be preserved (SPER).

If the prime rules are not applicable it will be useful to discard less than a half subinterval of the current interval $[S,T] \subseteq [\emptyset,N]$. In the following section we further relax most of the theoretical results presented in the previous sections of this chapter with the purpose to show the correctness of all excluding rules and their generalizations (preservation rules) based only on the definitions of submodularity and the maximum value $z^*[S,T]$ of the function z on the interval $[S,T] \subseteq [\emptyset,N]$.

2.4 Preservation Rules: Generalization and a Simple Justification

In the following theorem we give an important interpretation of the submodularity property which is based on two pairs of submodular function values. For this purpose we introduce an *upper (respectively, lower) partition* of the current interval $[S,T]$ by an inner vertex $Q : S \subset Q \subset T$ into two parts $[Q,T]$ and $[S,T] \setminus [Q,T]$

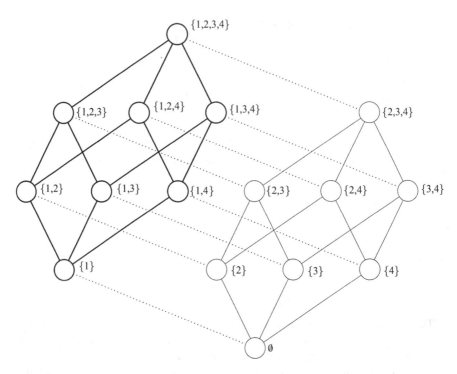

Fig. 2.8 Example of prime excluding rules

(respectively, $[S,Q]$ and $[S,T] \setminus [S,Q]$). In terms of the maximum values of the function z defined on each of two parts of the above-mentioned partitions, a special case of submodularity can be read as either $z^*([S,T] \setminus [Q,T]) + z(Q) \geq z(S) + z^*[Q,T]$ or $z^*([S,T] \setminus [S,Q]) + z(Q) \geq z^*[S,Q] + z(T)$.

Here, the both maximal values of a submodular function and their arguments (vertices) involved in each of the above indicated inequalities are *unknown*. In other words, Theorem 2.8 establishes a relationship of the difference between the unknown optimal values of z on the two parts of the partition, for example, $([S,T] \setminus [Q,T])$ and $[Q,T]$ of $[S,T]$ and the corresponding difference $z(S) - z(Q)$ (see the FER in Theorem 2.6); a symmetrical result is obtained for the SER.

Theorem 2.8. *Let z be a submodular function on the interval $[S,T] \subseteq [\emptyset,N]$. Then for any Q such that $S \subset Q \subset T$ the following assertions hold.*

(a) $z^*([S,T] \setminus [Q,T]) - z^*[Q,T] \geq z(S) - z(Q)$.
(b) $z^*([S,T] \setminus [S,Q]) - z^*[S,Q] \geq z(T) - z(Q)$.

Proof. We prove only case (a) because the proof of case (b) is similar. Let $z^*[Q,T] = z(Q \cup J)$ with $J \subseteq T \setminus Q$. Define $I = S \cup J$. Then $I \in [S,T] \setminus [Q,T]$ since $Q \setminus S \not\subseteq I$. We have that $z^*([S,T] \setminus [Q,T]) - z(S) \geq z(I) - z(S) = z(S \cup J) - z(S)$. From the submodularity of z we have $z(S \cup J) - z(S) \geq z(Q \cup J) - z(Q)$. Therefore, $z^*([S,T] \setminus [Q,T]) - z(S) \geq z^*[Q,T] - z(Q)$. □

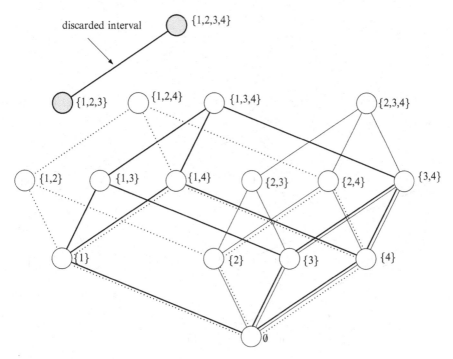

Fig. 2.9 A representation of the upper partition of the interval $[S,T] = [\emptyset, \{1,2,3,4\}]$ with $Q \setminus S = \{1,2,3\}$.

Theorem 2.8 is a generalization of Cherenin–Khachaturov's rules stating that the difference of values of a submodular function on any pair of embedded subsets is a lower bound for the difference between the optimal values of z on the two parts of the partition defined by this pair of embedded subsets. The theorem can be used to decide in which part of the partition $([S,T] \setminus [Q,T])$ and $[Q,T]$ of $[S,T]$ a global maximum of z is located.

We may represent the partition of interval $[S,T]$ from Theorem 2.8 by means of its proper subintervals as follows:

$$(a)\ upper\ partition\ [S,T] \setminus [Q,T] = \bigcup_{i \in Q \setminus S} [S, T-i]$$

and

$$(b)\ lower\ partition\ [S,T] \setminus [S,Q] = \bigcup_{i \in T \setminus Q} [S+i, T].$$

Examples of upper and lower partitions in Figs. 2.9 and 2.10 are shown.

A disadvantage of representations (a) and (b) is a nonempty overlapping of each pairwise distinct intervals involved in these representations. As easy to see in Figs. 2.11 and 2.12 we can avoid such an overlapping by representing the remaining

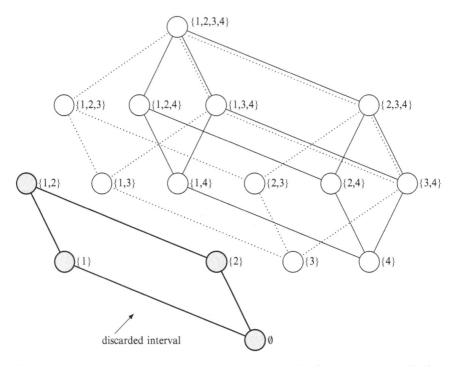

Fig. 2.10 A representation of the lower partition by $Q = \{1,2\}$ for the interval $[S,T] = [\emptyset, \{1,2,3,4\}]$ with $T \setminus Q = \{3,4\}$

parts $([S,T] \setminus [Q,T])$ and $([S,T] \setminus [S,Q])$ with a sequence of "parallel" nonoverlapping intervals. For example, the difference $[\emptyset, \{1,2,3,4\}] \setminus [\{1,2,3\},\{1,2,3,4\}] = [\{1,2\},\{1,2,4\}] \cup [\{1\},\{1,3,4\}] \cup [\emptyset,\{2,3,4\}]$ (see Figs. 2.9 and 2.10), and the difference $[\emptyset, \{1,2,3,4\}] \setminus [\emptyset,\{1,2\}] = [\{3\},\{1,2,3\}] \cup [\{4\},\{1,2,3,4\}]$ (see Figs. 2.11 and 2.12).

The sequence of nonoverlapping intervals can be created by the following iterative procedure. We will use the value $d = \dim([U,W])$ of the *dimension* of an interval $[U,W]$ interpreted as the corresponding subspace of the Boolean space $\{0,1\}^n$ which is another representation of the interval $[\emptyset, N]$.

If we have discard the k-dimensional subinterval $[Q,T]$ in the upper partition of the interval $[S,T]$, then the first nonoverlapping interval $[U_1,W_1]$ is the k-dimensional subinterval of the $(k+1)$-dimensional interval $[U_1,T] = [Q,T] \cup [U_1,W_1]$. In other words, the first nonoverlapping interval $[U_1,W_1]$ is the k-dimensional complement to the $(k+1)$-dimensional interval $[U_1,T]$ such that $[U_1,W_1] = [U_1,T] \setminus [Q,T]$. The second nonoverlapping interval $[U_2,W_2]$ is the $(k+1)$-dimensional subinterval of the $(k+2)$-dimensional interval $[U_2,T] = [U_1,T] \cup [U_2,W_2]$, and $[U_2,W_2] = [U_2,T] \setminus [U_1,T]$, etc. Finally, $[U_q,W_q] = [U_q,T] \setminus [U_{(q-1)},T]$. The number q of the nonoverlapping intervals in the upper partition is equal to $n - k$, where $k = \dim[Q,T]$. The representation of a lower partition by the sequence of nonoverlapping intervals

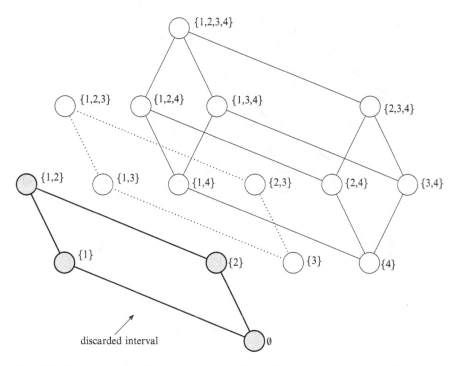

Fig. 2.11 The non-overlapping representation of the lower partition by parallel intervals $[\{3\},\{1,2,3\}]$ and $[\{4\},\{1,2,3,4\}]$.

can be described in similar lines. Note that the above indicated representation of lower (upper) partition by a sequence of nonoverlapping intervals has the *minimum number of mutually disjoint intervals*.

For example (see Fig. 2.12), the complement interval to $[\{1,2,3\},\{1,2,3,4\}]$ is $[\{1,2\},\{1,2,4\}]$ since $[\{1,2\},\{1,2,4\}]\cup[\{1,2,3\},\{1,2,3,4\}]=[\{1,2\},\{1,2,3,4\}]$, and the complement to $[\{1,2\},\{1,2,3,4\}]$ is $[\{1\},\{1,3,4\}]$. Finally, the complement to $[\{1\},\{1,2,3,4\}]$ is $[\emptyset,\{2,3,4\}]$.

If, in Theorem 2.8, we replace Q by $S+k$ in part (a), and Q by $T-k$ in part (b), we obtain the following generalization of the prime rules stated in Theorem 2.7.

Corollary 2.2. *Let z be a submodular function on the interval $[S,T]\subseteq[\emptyset,N]$ and let $k\in T\setminus S$. Then the following assertions hold.*

(a) $z^*[S,T-k]-z^*[S+k,T]\geq z(S)-z(S+k)$.
(b) $z^*[S+k,T]-z^*[S,T-k]\geq z(T)-z(T-k)$.

By adding the condition $z(S)-z(S+k)\geq 0$ to part (a) and the condition $z(T)-z(T-k)\geq 0$ to part (b) of Corollary 2.2, we obtain another form (see Corollary 2.3) of two prime rules from Theorem 2.7 for preserving subintervals containing at least one global maximum of z on $[S,T]$.

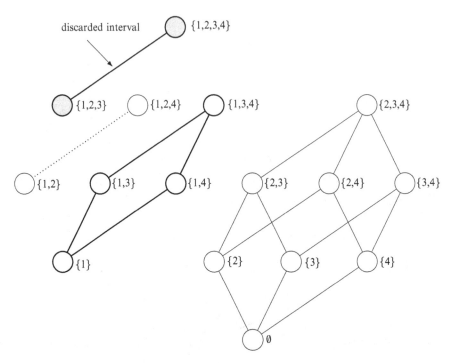

Fig. 2.12 The nonoverlapping representation of the upper partition by the parallel intervals $[\{1,2\},\{1,2,4\}]$, $[\{1\},\{1,3,4\}]$, and $[\emptyset,\{2,3,4\}]$

Corollary 2.3. *Let z be a submodular function on the interval $[S,T] \subseteq [\emptyset,N]$ and $k \in T \setminus S$. Then the following assertions hold.*

(a) *First Preservation (FP) Rule.*
    If $z(S) \geq z(S+k)$, then $z^*[S,T] = z^*[S,T-k] \geq z^*[S+k,T]$.
(b) *Second Preservation (SP) Rule.*
    If $z(T) \geq z(T-k)$, then $z^*[S,T] = z^*[S+k,T] \geq z^*[S,T-k]$.

Proof. (a) From Corollary 2.2a we have $z^*[S,T-k] - z^*[S+k,T] \geq z(S) - z(S+k)$.
    By assumption $z(S) - z(S+k) \geq 0$. Hence, $z^*[S,T] = z^*[S,T-k] \geq z^*[S+k,T]$.
(b) The proof is similar.   □

From the calculation point of view these rules are the same as in Theorem 2.6, but Theorem 2.7 is more powerful than Corollary 2.3. In Theorem 2.7 we preserve at least one strict local maximum from each STC, and hence one global maximum from each STC that contains global maxima. Corollary 2.3 only states that we preserve at least one global maximum. However, we can use Corollary 2.3 for constructing some extension of the preservation rules.

For $\varepsilon \geq 0$, we may consider the problem of finding an approximate solution $J \in [S,T]$ such that $z^*[S,T] \leq z(J) + \varepsilon$; J is called an ε-*maximum* of z on $[S,T]$.

The following corollary presents an extension of the rules from Corollary 2.3 which is appropriate to the problem of ε-maximization.

Corollary 2.4. *Let z be a submodular function on the interval $[S,T] \subseteq [0,N]$, and $k \in T \setminus S$. Then the following assertions hold.*

(a) *First θ-Preservation (θ-FP) Rule.*
   If $z(S) - z(S+k) = \theta < 0$, then $z^*[S,T] - z^*[S,T-k] \leq -\theta$, which means that $[S,T-k]$ contains a $|\theta|$-maximum of $[S,T]$.
(b) *Second η-Preservation (η-SP) Rule.*
   If $z(T) - z(T-k) = \eta < 0$, then $z^*[S,T] - z^*[S+k,T] \leq -\eta$, which means that $[S+k,T]$ contains a $|\eta|$-maximum of $[S,T]$.

Proof. The proof of part (a) is as follows.

Case 1. If $z^*[S,T] = z^*[S,T-k]$ then $z^*[S,T-k] - z^*[S,T-k] \leq -\theta$ or $z^*[S,T] - z^*[S,T-k] \leq -\theta$.

Case 2. If $z^*[S,T] = z^*[S+k,T]$, then from Theorem 2.7a follows that $z^*[S,T-k] - z^*[S+k,T] \geq \theta$ or $z^*[S,T-k] - z^*[S,T] \geq \theta$. Hence $z^*[S,T] - z^*[S,T-k] \leq -\theta$. The proof of (b) is similar.   □

2.5   The PPA

By means of Corollary 2.3 it is often possible to exclude a large part of $[0,N]$ from consideration when determining a global maximum of z on $[0,N]$. The so-called *PPA* (see [66]) determines the smallest subinterval $[S,T]$ of $[0,N]$ containing a global maximum of z, by using the preservation rules of Corollary 2.3.

We call the PPA the *dichotomy algorithm* because in every successful step it halves the current domain of a submodular function.

Let $[S,T]$ be an interval. For each $i \in T \setminus S$, define $\delta^+(S,T,i) = z(T) - z(T-i)$ and $\delta^-(S,T,i) = z(S) - z(S+i)$; moreover, define $\delta^+_{max}(S,T) = \max\{\delta^+(S,T,i) \mid i \in T \setminus S\}$, $r^+(S,T) = \min\{r \mid \delta^+(S,T,r) = \delta^+_{max}(S,T)\}$. Similarly, for $\delta^-(S,T,i))$ define $\delta^-_{max}(S,T) = \max\{\delta^-(S,T,i)) \mid i \in T \setminus S\}$, $r^-(S,T) = \min\{r \mid \delta^-(S,T,r) = \delta^-_{max}(S,T)\}$. If no confusion is likely, we briefly write r^-, r^+, δ^-, δ^+ instead of $r^-(S,T)$, $r^+(S,T)$, $\delta^-_{max}(S,T)$, and $\delta^+_{max}(S,T)$ respectively (Fig. 2.13).

Each time that either S or T are updated during the execution of the PPA, the conditions of Corollary 2.3 remain satisfied, and therefore $z^*[S,T] = z^*[U,W]$ with $[U,W] \subseteq [S,T]$ remains invariant at each step of the PPA. At the end of the algorithm we have that $\max\{\delta^+, \delta^-\} < 0$, and therefore $z(S) < z(S+i)$ and $z(T) < z(T-i)$ for each $i \in T \setminus S$. Hence Corollary 2.3 cannot be applied to further reduce the interval $[S,T]$ without violating $z^*[S,T] = z^*[U,W]$. Note that this remark shows the correctness of the procedure PP(.).

If we replace in the PPA the rules of Corollary 2.3 by those of Corollary 2.4, we obtain an ε-maximization variant of the PPA. In this case the output of the

Procedure $PP(U,W,S,T)$
Input: A submodular function z on the subinterval
       $[U,W]$ of $[0,N]$
Output: A subinterval $[S,T]$ of $[U,W]$ such that
       $z^*[S,T] = z^*[U,W]$, $z(S) < z(S+i)$ and
       $z(T) < z(T-i)$ for each $i \in T \setminus S$.
begin
        $S \leftarrow U;$ $T \leftarrow W;$
        Step 1: if $S = T$
           then goto Step 4;
        Step 2: Calculate δ^+ and r^+;
           if $\delta^+ \geq 0$ (Corollary 2.3b)
           then begin call $PP(S+r^+,T;S,T)$
                 goto Step 4
           end;
        Step 3: Calculate δ^- and r^-;
           if $\delta^- \geq 0$ (Corollary 2.3a)
           then begin call $PP(S,T-r^-;S,T)$
                 goto Step 4
           end;
        Step 4:
end;

Fig. 2.13 The dichotomy (preliminary preservation) algorithm

ε-PPA will be a subinterval $[S,T]$ of $[U,W]$ such that $z^*[U,W] - z^*[S,T] \leq \varepsilon$ with postconditions $z(S) + \varepsilon < z(S+i)$ and $z(T) + \varepsilon < z(T-i)$ for each $i \in T \setminus S$.

The following theorem can also be found in [58, 66]. It provides an upper bound for the worst case complexity of the PPA; the complexity function is dependent only on the number of comparisons of pairs of values for z.

Theorem 2.9. *The time complexity of the PP algorithm procedure is at most* $O(n^2)$.

Note that if the PPA terminates with $S = T$, then S is a global maximum of z. Any submodular function z on $[U,W]$ for which the PP algorithm returns a global maximum for z is called a *PP-function*.

An example of a set of PP-functions $\mathscr{P}$ is shown in Fig. 2.14. Here, for all vertices without prespecified values of $z(I)$ can be assigned an arbitrary value of $z(I)$ such that each corresponding set function $z(I) \in \mathscr{P}$ defined on the whole weighted graph G will be submodular. For example, if for all vertices without prespecified values of $z(I)$ in Fig. 2.14 we set $z(I) = a$, then for each real valued constant $a : 2 \leq a \leq 3$ the corresponding function z is a submodular PP-function. It means that by applying the dichotomy algorithm we have found an optimal solution to the PSMF for all PP-functions defined by a constant a.

Corollary 2.5 describes in terms of STCs some properties of the variables S and T during the iterations of the PPA. A representative $L_1^j \in V_0^j$ with $j \in J_1$ which will be preserved through all iterations during the execution of the PPA by FPER

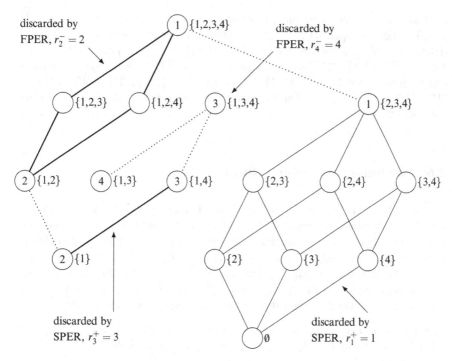

Fig. 2.14 The idea of the dichotomy algorithm: $z(\{1,3\}) = 4$ is the global maximum for all submodular functions from the subclass of $\mathscr{P}$

$(L_1^j \in V_0^j \cap [S, T - i] \neq \emptyset$ with $j \in J_1)$ or SPER $(L_1^j \in V_0^j \cap [S + i, T] \neq \emptyset$ with $j \in J_1)$ is called a *PP-representative* of STC H_0^j with $j \in J_1$ (see Theorem 2.7).

Corollary 2.5. *If z is a submodular PP-function on $[U, W] \subseteq [\emptyset, N]$, then at each iteration of the PPA $S \subseteq \cap_{j \in J_1} L_1^j$ and $T \supseteq \cup_{j \in J_1} L_1^j$.*

Proof. Theorem 2.7a says that if $z(S + i) - z(S) \leq 0$ for some $i \in T \setminus S$, then by preserving the interval $[S, T - i]$ we preserve at least one PP-representative L_1^j from each STC H_0^j, and hence $i \notin L_1^j$. In case of Theorem 2.7b we preserve PP-representatives L_1^j such that $i \in L_1^j$ for all STCs in $[S, T]$. Therefore, $i \in S \subseteq \cap_{j \in J_1} L_1^j$ and $T \supseteq \cup_{j \in J_1} L_1^j$. ∎

The following theorem gives a property of PP-functions in terms of STCs.

Theorem 2.10. *If z is a submodular PP-function on $[U, W] \subseteq [\emptyset, N]$, then $[U, W]$ contains exactly one STC.*

Proof. From $\cap_{j \in J_1} L_1^j \supseteq S = T \supseteq \cup_{j \in J_1} L_1^j$ we obtain $\cap_{j \in J_1} L_1^j = \cup_{j \in J_1} L_1^j$ or $L_1^j = L$ for all $j \in J_1$. ∎

Note that not each submodular function with exactly one STC on $[\emptyset, N]$ is a PP-function. For example, let $N = \{1,2,3\}$ and consider the submodular function z defined by $z(I) = 2$ for any $I \in [\emptyset, \{1,2,3\}] \setminus (\{\emptyset\} \cup \{1,2,3\})$ and $z(I) = 1$ for $I \in (\{\emptyset\} \cup \{1,2,3\})$. The vertex set of the unique STC defined by this function can be represented by $[\{1\},\{1,2\}] \cup [\{1\},\{1,3\}] \cup [\{2\},\{1,2\}] \cup [\{2\},\{2,3\}] \cup [\{3\},\{1,3\}] \cup [\{3\},\{2,3\}]$. The PPA terminates with $[S,T] = [\emptyset, \{1,2,3\}]$ and so, z is not a PP-function.

2.6　Nonbinary Branching Rules

Usually in BnB type algorithms we use a *binary* branching rule by which the original set $[S,T]$ of feasible solutions will be split by an element k into two subsets $[S+k,T]$ and $[S,T-k]$. Let us consider an interval $[S,T]$ for which the postconditions of the PPA are satisfied, i.e., $z(S) < z(S+i)$ and $z(T) < z(T-i)$ for each $i \in T \setminus S$. Thus the PPA cannot make the interval $[S,T]$ smaller. By using Corollary 2.6 we can sometimes find two subintervals $[S,T-k_1]$ and $[S,T-k_2]$ such that the postconditions of the PPA algorithm for each of these intervals are violated.

Corollary 2.6. *Let z be a submodular function on the interval $[S,T] \subseteq [\emptyset, N]$ and let $k_1, k_2 \in T \setminus S$ with $k_1 \neq k_2$. Then the following assertions hold.*

(a) $\max\{z^*[S,T-k_1], z^*[S,T-k_2]\} - z^*[S+k_1+k_2,T] \geq z(S) - z(S+k_1+k_2)$.
(b) $\max\{z^*[S+k_1,T], z^*[S+k_2,T]\} - z^*[S,T\setminus\{k_1,k_2\}] \geq z(T) - z(T\setminus\{k_1,k_2\})$.

Proof. We prove only part (a) because the proof of part (b) is similar. Replace Q by $S+k_1+k_2$ in Theorem 2.8a. Then, $z^*([S,T]\setminus[Q,T]) - z^*[Q,T] = z^*(\bigcup_{i\in Q\setminus S}[S,T-i]) - z^*[Q,T] = z^*([S,T-k_1] \cup [S,T-k_2]) - z^*[S+k_1+k_2,T] = \max\{z^*[S,T-k_1], z^*[S,T-k_2]\} - z^*[S+k_1+k_2,T] \geq z(S) - z(Q) = z(S) - z(S+k_1+k_2)$. □

In the case that $z(S) - z(S+k_1+k_2) \geq 0$ we can discard the interval $[S+k_1+k_2,T]$ and continue the search for an optimal solution by applying the PPA separately to each remaining interval $[S,T-k_1]$ and $[S,T-k_2]$, which are obtained by subtracting an element k_i from T. The symmetrical case will be obtained if $z(T) - z(T\setminus\{k_1,k_2\}) \geq 0$. Corollary 2.6 can easily be generalized to the case of m-ary branching by elements $k_1,k_2,\ldots,k_m$ with $m \leq |T \setminus S|$.

We conclude this section with a simple plant location example borrowed from [18]. The data are presented in Table 2.1.

For solving the SPLP it suffices to solve the problem $\min\{z(I) \mid I \in [\emptyset, N]\} = z^*[\emptyset, N] = z(G)$ with $N = \{1,2,3,4\}$, $m = 5$ and

$$z(I) = \sum_{i\in I} f_i + \sum_{j=1}^{m} \min_{i\in I} c_{ij}.$$

Table 2.1 The data of the SPLP

Location		Delivery cost to site				
i	f_i	$j=1$	$j=2$	$j=3$	$j=4$	$j=5$
1	7	7	15	10	7	10
2	3	10	17	4	11	22
3	3	16	7	6	18	14
4	6	11	7	6	12	8

As usual for the SPLP, f_i is the fixed cost of opening a plant at location i, c_{ij} is the cost of satisfying the demand of customer j by plant i, and $z(I)$ is a supermodular function. Note that if in the definition of a submodular function we change the sign "$\geq$" to the opposite sign "$\leq$," then we obtain the definition of a supermodular function. For the sake of completeness, let us show that $z(I)$ of the SPLP is supermodular

Lemma 2.3. *The objective $z(I)$ of the SPLP is supermodular.*

Proof. According to Theorem 2.1(i) a function is supermodular if

$$z(A) + z(B) \leq z(A \cup B) + z(A \cap B), \quad \forall A, B \subseteq N.$$

We use the following representation of this definition

$$z(A) + z(B) - z(A \cup B) + z(A \cap B) \leq 0, \quad \forall A, B \subseteq N.$$

Substituting

$$z(I) = \sum_{i \in I} f_i + \sum_{j=1}^{m} \min_{i \in I} c_{ij}$$

gives

$$\sum_{i \in A} f_i + \sum_{j=1}^{m} \min_{i \in A} c_{ij} + \sum_{i \in B} f_i + \sum_{j=1}^{m} \min_{i \in B} c_{ij}$$
$$- \sum_{i \in A \cup B} f_i - \sum_{j=1}^{m} \min_{i \in A \cup B} c_{ij} - \sum_{i \in A \cap B} f_i - \sum_{j=1}^{m} \min_{i \in A \cap B} c_{ij}$$
$$= \sum_{i \in A} f_i + \sum_{i \in B} f_i - \sum_{i \in A \cup B} f_i - \sum_{i \in A \cap B} f_i$$
$$+ \sum_{j=1}^{m} [(\min_{i \in A} c_{ij} - \min_{i \in A \cup B} c_{ij}) + (\min_{i \in B} c_{ij} - \min_{i \in A \cap B} c_{ij})].$$

Note that

$$\left[\sum_{i \in A} f_i + \sum_{i \in B} f_i - \sum_{i \in A \cup B} f_i - \sum_{i \in A \cap B} f_i \right] = 0,$$

hence it is enough to show that for each $j = 1, \ldots, m$

$$[(\min_{i \in A} c_{ij} - \min_{i \in A \cup B} c_{ij}) + (\min_{i \in B} c_{ij} - \min_{i \in A \cap B} c_{ij})] \leq 0.$$

Let us consider two cases.

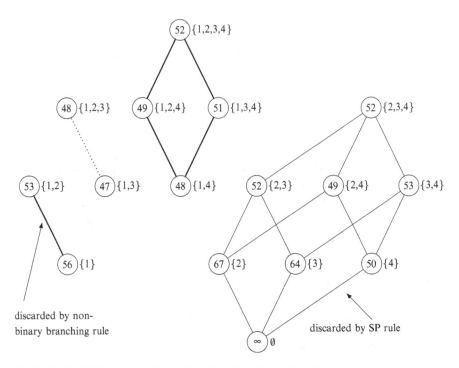

Fig. 2.15 The SPLP example: illustration of nonbinary branching rule

Case 1. $\min_{i \in A \cup B} c_{ij} = c_{aj}$ for some $a \in A$. Then $\min_{i \in A} c_{ij} = \min_{i \in A \cup B} c_{ij}$ and $\min_{i \in B} c_{ij} \le \min_{i \in A \cap B} c_{ij}$.

Case 2. $\min_{i \in A \cup B} c_{ij} = c_{bj}$ for some $b \in B$. Then $\min_{i \in B} c_{ij} = \min_{i \in A \cup B} c_{ij}$ and $\min_{i \in A} c_{ij} \le \min_{i \in A \cap B} c_{ij}$.   □

We use this example for illustrating that the supermodular function defined by data from Table 2.1 is not a PP-function. Of course, here we mean the corresponding definition of a PP-function obtained by replacing the definitions of local, global maxima of a submodular function by the local, global minima of a supermodular function. It is easy to check that this supermodular function has two trivial analogues of STCs: $\{1,4\}, \{1,3\}$ and one trivial analogue of SDC: $\{2,4\}$ (see Fig. 2.15).

After the first execution of Step 3 of the PPA, we have that $[S,T] = [\{1\}, \{1, 2, 3, 4\}]$, because $\delta^+ = z(\{1,2,3,4\}) - z(\{2,3,4\}) = 0$ and $r^+ = 1$. Together with interval $[\{\emptyset\}, \{2,3,4\}]$ the PPA has discarded the trivial SDC $\{2,4\}$. After the second execution of Steps 2 and 3 the PPA terminates with interval $[S,T] = [\{1\}, \{1,2,3,4\}]$, because all postconditions of the PPA are satisfied. Hence, this function is not a PP-function. A global minimum of this SPLP can be found by application of the following analogue of the inequality from Corollary 2.6b:

$$\min\{z^*[S+k_1,T], z^*[S+k_2,T]\} - z^*[S,T \setminus \{k_1,k_2\}]$$
$$\le z(T) - z(T \setminus \{k_1,k_2\}).$$

Let us substitute all possible pairs $\{k_1, k_2\}$ into the right-hand side of this inequality with $S = \{1\}$ and $T = \{1,2,3,4\}$. Then, we have that only $z(\{1,2,3,4\}) - z(\{1,2,3,4\} - \{3,4\}) = 52\text{-}53 < 0$. Hence, we can discard the interval $[\{1\}, \{1,2,3,4\} - \{3,4\}]$ and we may continue to find $z^*[\{1\}, \{1,2,3,4\}]$ by solving two remaining subproblems $z^*[\{1,3\}, \{1,2,3\}]$ and $z^*[\{1,4\}, \{1,2,3,4\}]$ defined on "parallel" intervals $[\{1,3\}, \{1,2,3\}]$ and $[\{1,4\}, \{1,2,3,4\}]$ (with disjoint set of feasible solutions) instead of two corresponding subproblems $z^*[S + k_1, T] = z^*[\{1,3\}, \{1,2,3,4\}]$ and $z^*[S + k_2, T] = z^*[\{1,4\}, \{1,2,3,4\}]$ which have the nonempty intersection on $[\{1,3,4\}, \{1,2,3,4\}]$. Each of these subproblems can be solved by the corresponding analogue of the PPA.

2.7 Concluding Remarks

We have considered a submodular function z defined on the Boolean hypercube to which we can apply a classic theorem of Cherenin that z is quasiconcave on any chain that intersects a local maxima component. This result enables a clearer understanding of the structure of a submodular function in terms of components of the graph of local maxima. Specifically we may state that each component of the graph of local maxima is a maximal connected set of intervals whose end points are lower and upper local maxima. Cherenin's theorem provides a justification of "the method of successive calculations." This method was successfully applied to solve problems arising in railway logistics planning (see [27, 28, 118]), and for constructing BnB type algorithms (see [3,46,56,57,60–64,66,72,89,90]) for solving a number of NP-hard problems.

We have shown that if the dichotomy algorithm (PPA) terminates with $S = T$, then the given submodular function has exactly one strict component of local maxima (STC). Hence the number of subproblems created in a branch without bounds type algorithm, which is based on the dichotomy algorithm, can be used as an upper bound for the number of the STCs. In a similar way, an upper bound for the number of all components (STCs and SDCs) by using strict excluding rules can be calculated. This information can be used for complexity analysis in terms of the number of local optima for a specific class of problems arisen in practice (computational experiments).

We next proposed a generalization of Cherenin's excluding rules given by Theorem 2.8 which provides implicit enumeration bounds for a recursive implementation of any BnB procedure incorporating the dichotomy algorithm. This generalization is useful in two respects. First, it is suitable for use in ε-*optimal* procedures which obtain an approximate global maximum within specified bounds. Second, the theorem allows the derivation of alternatives to the prime excluding rules by which we are able to discard subintervals of smaller cardinality than half original subinterval. We show that the remaining part of the current interval can be represented by a set of subintervals, some of which may include just one strict component. In other words, we try to prepare the necessary conditions for the dichotomy algorithm to terminate on each subinterval. Moreover, Theorem 2.8 is based only on the definition of the maximum value of PMSF for an interval of $[0, N]$, and relaxed Cherenin–Khachaturov's theory presented in Sects. 2.2 and 2.3 (which is based on notions of monotonicities on a chain, local and global maxima, strict and saddle components in the Hasse diagram).

Corollary 2.2 can be considered as the basis of our data correcting (DC) algorithm presented in the next chapter. It states that if an interval $[S, T]$ is split into $[S, T - k]$ and $[S + k, T]$, then the difference between the submodular function values $z(S)$ and $z(S + k)$, or between the values of $z(T)$ and $z(T - k)$ is an upper bound for the difference of the (unknown!) optimal values on the two subintervals. This difference is used for "correcting" the current data (values of a submodular function z) in the DC algorithm. In the next chapter our computational experiments with the QCP problem show that we can substantially reduce the calculation time for data correcting algorithms [59,66] by recursive application of our main theorem.

An interesting subject for future research is the investigation of the computational efficiency of m-ary branching rules (see Corollary 2.6) for specific problems which can be reduced to the maximization of submodular functions.

Another purpose of this chapter is to present our main result [73] which can be stated as follows. For any partition of the current Hasse subdiagram $[S, T]$ spanned on a pair of embedded subsets $\emptyset \subseteq S \subset Q \subset T \subseteq N = \{1, 2, \ldots, n\}$ into two parts either $[S, T] \setminus [Q, T]$ and $[Q, T]$ or $[S, T] \setminus [S, Q]$ and $[S, Q]$ defined by an inner subset Q from this subdiagram, the difference of two corresponding function values either $z(S) - z(Q)$ or $z(T) - z(Q)$ is a lower bound for the difference between the unknown(!) optimal values either $z^*([S, T] \setminus [Q, T]) - z^*([Q, T])$ or $z^*([S, T] \setminus [S, Q]) - z^*([S, Q])$, respectively, of the submodular function z. The main result was successfully used as a base of Data Correcting (DC) algorithms for the maximization of general submodular functions and its special cases, for example, the QCP and simple plant location problems (which is a special case of minimization a supermodular function). Cherenin's Excluding Rules, the dichotomy algorithm, and its generalization with the new branching rules are easy corollaries of our main result. The usefulness of our new branching rules is illustrated by means of a numerical Simple Plant Location Problem example.

Chapter 3
Data Correcting Approach for the Maximization of Submodular Functions

The data correcting (DC) algorithm is a recursive Branch-and-Bound (BnB) type algorithm, in which the data of a given problem instance are "heuristically corrected" at each branching in such a way that the new instance will be as close as possible to polynomially solvable and the result satisfies a prescribed accuracy (the difference between optimal and current solution). The main idea of the data correcting approach for an arbitrary function z defined on a set S can be described as follows (see e.g., [66]). Let y belongs to a polynomially solvable class of functions which is a subclass of a given class of functions z defined also on S, and let

$$\rho(z,y) = \max\{|z(s) - y(s)| \ : \ s \in S\}$$

be the proximity measure. If we denote the maximal values of z and y on S by $z^*(S) = z(s_z)$ and $y^*(S) = y(s_y)$, then an analogue of Theorem 1.2 is read as follows:

Theorem 3.1.
$$|z^*(S) - y^*(S)| \leq \rho(z,y).$$

Proof. If $z^*(S) \geq y^*(S)$ then $z^*(S) - y^*(S) = |z^*(S) - y^*(S)| = z(s_z) - y(s_y) \leq z(s_z) - y(s_z) = |z(s_z) - y(s_z)| \leq \rho(z,y)$. Otherwise, $z^*(S) < y^*(S)$, we have $y^*(S) - z^*(S) = |y^*(S) - z^*(S)| = |z^*(S) - y^*(S)| = y(s_y) - z(s_z) \leq y(s_y) - z(s_y) = |y(s_y) - z(s_y)| = |z(s_y) - y(s_y)| \leq \rho(z,y)$. $\square$

Let us remind that we assume an efficient (polynomial) computing of the value of $\rho(z,y)$. In general if the problem of finding $z^*(S)$ is intractable, then the computing of $\rho(z,y)$ is also intractable. In this chapter we replace the computing of intractable value of $\rho(z,y)$ by computing an upper bound of $\rho(z,y)$ which is tractable and based on the so-called *correcting rules*.

In this chapter the data correcting (DC) algorithm is applied to determining exact or approximate global maxima (respectively, minima) of submodular (respectively, supermodular) functions. The algorithm is illustrated by an instance of the

B. Goldengorin and P.M. Pardalos, *Data Correcting Approaches in Combinatorial Optimization*, SpringerBriefs in Optimization, DOI 10.1007/978-1-4614-5286-7_3,
© Boris Goldengorin, Panos M. Pardalos 2012

simple plant location problem (SPLP). The computational results, obtained for the quadratic cost partition (QCP) problem, show that the DC algorithm outperforms a branch-and-cut algorithm, not only for sparse graphs but also for nonsparse graphs (with density more than 40%) often with speeds 100 times faster.

3.1   The Main Idea of the Data Correcting Algorithm: An Extension of the PPA

Recall that if a submodular function z is not a PP-function, then the PP algorithm terminates with a subinterval $[S,T]$ of $[\emptyset,N]$ with $S \neq T$ that contains a maximum of z without knowing its exact location in $[S,T]$. In this case, the conditions

$$\delta^+ < 0 \ \text{ for } i \in T \backslash S, \tag{3.1}$$

$$\delta^- < 0 \ \text{ for } i \in T \backslash S, \tag{3.2}$$

are satisfied at termination of the PP algorithm. The basic idea of the DC algorithm is that if a situation occurs for which both (3.1) and (3.2) hold, then the data of the current problem will be "corrected" in such a way that a *corrected* function y violates at least one of the conditions (3.1) or (3.2).

In this section we will restrict ourselves to the situation for which the submodularity of the corrected function is easy to prove (see [66]). Hence a situation is studied for which there is an element $i \in T \backslash S$ such that, either $y(T - i) \leq y(T)$, or $y(S + i) \leq y(S)$ holds. Now Corollary 2.3 can be applied again, and we are in the situation that the PP algorithm can be applied. For all possible elements i we try to choose one for which the correction procedure maintains a solution within the prescribed bound ε_0. If such an element i does not exist, we choose an arbitrary $i \in T \backslash S$ and branch the current problem into two subproblems, one on $[S+i,T]$ and one on $[S, T - i]$. We should in any case find answers to the following questions:

- How should the difference between the values of a global maximum of the corrected and the uncorrected functions be estimated, and, how does this difference depend on the specific corrections?
- How should the above-mentioned difference be decreased in case it does not satisfy the prescribed accuracy ε_0?

The answers to these questions can be found below. In order to preserve the submodularity we will use the following correcting rules.

Let $\emptyset \subseteq S \subseteq T \subseteq N$, and $r^+, r^- \in T \backslash S$. Moreover, let y be a submodular function on $[\emptyset,N]$. For each $I \in [S,T]$ define the following two correcting rules.

Correcting Rule 1:

$$y(I) = \begin{cases} z(I) + \delta^+, & \text{if } I \in [S, T - r^+]; \\ z(I), & \text{otherwise} \end{cases}$$

Correcting Rule 2:

$$y(I) = \begin{cases} z(I) + \delta^-, & \text{if } I \in [S + r^-, T]; \\ z(S), & \text{otherwise} \end{cases}$$

It can be easily seen that if z is submodular on a certain interval, then so is y.

An extension of the PP algorithm is based on the statements of the following lemma.

Lemma 3.1. *Let z be a submodular function on the interval $[S, T] \subseteq [\emptyset, N]$ and let $i \in T \setminus S$. Then*

(a) *If $\delta^- = z(S) - z(S + i) \geq 0$ and $z^*[S, T - i] - z(\lambda) \leq \gamma \leq \varepsilon$, then $z^*[S, T] - z(\lambda) \leq \gamma \leq \varepsilon$.*

(b) *If $\delta^+ = z(T) - z(T - i) \geq 0$ and $z^*[S + i, T] - z(\lambda) \leq \gamma \leq \varepsilon$, then $z^*[S, T] - z(\lambda) \leq \gamma \leq \varepsilon$.*

(c) *If $-\varepsilon \leq \delta^- = z(S) - z(S + i) < 0$ and $z^*[S, T - i] - z(\lambda) \leq \gamma \leq \varepsilon + \delta^-$, then $z*[S, T] - z(\lambda) \leq \gamma - \delta^- \leq \varepsilon$.*

(d) *If $-\varepsilon \leq \delta^+ = (T) - z(T - i) < 0$, and $z^*[S + i, T] - z(\lambda) \leq \gamma \leq \varepsilon + \delta^+$, then $z^*[S, T] - z(\lambda) \leq \gamma - \delta^+ \leq \varepsilon$.*

Proof. The proof of (a) is as follows. From $\delta^- \geq 0$ and Corollary 2.3a we obtain $z^*[S, T] = z^*[S, T - i]$. Hence $z^*[S, T] - z^*(\lambda) = z^*[S, T - i] - z^*(\lambda) \leq \gamma \leq \varepsilon$. Since the proof of (b) is similar to that of (a) we conclude with a proof of (c).

From $\delta^- < 0$ and Corollary 2.4a we obtain $z^*[S, T] - z^*[S, T - i] \leq \delta^-$ or $z^*[S, T] \leq z^*[S, T - i] - \delta^-$ or $z*[S, T] - z(\lambda) \leq z^*[S, T - i] - \delta^- - z^*(\lambda) \leq \gamma - \delta^- \leq \varepsilon + \delta^- - \delta^- = \varepsilon$. □

The following theorem defines the branching step and shows how a current value of γ can be decreased.

Theorem 3.2. *Let z be an arbitrary function defined on a finite set S and let $S = \cup_{t \in P} S_t$ with $\max\{z(\lambda) \mid \lambda \in S_t\} = z^*(S_t)$, for $t \in P = \{1, \ldots, p\}$. Then for any $\varepsilon \geq 0$ the following assertion holds.*
If $z^*(S_t) - z(\lambda_t) \leq \gamma_t \leq \varepsilon$ for some $\lambda_t \in S$ and for all $t \in P$,
then $z^*(S) - \max\{z(\lambda_t) \mid t \in P\} \leq \max\{z(\lambda_t) + \gamma_t \mid t \in P\} - \max\{z(\lambda_t) \mid t \in P\} = \gamma \leq \max\{\gamma_t \mid t \in P\} \leq \varepsilon$.

Proof. $z^*(S) - \max\{z(\lambda_t) \mid t \in P\} = \max\{z^*(S_t) \mid t \in P\} - \max\{z(\lambda_t) \mid t \in P\} \leq \max\{z(\lambda_t) + \gamma_t \mid t \in P\} - \max\{z(\lambda_t) \mid t \in P\} = \gamma \leq \max\{z(\lambda_t) \mid t \in P\} + \max\{\gamma_t \mid t \in P\} - \max\{z(\lambda_t) \mid t \in P\} = \max\{\gamma_t \mid t \in P\} \leq \varepsilon$. □

Note that z need not be submodular in Theorem 3.2. It is clear from the proof of Theorem 3.2 that γ is independent on the order in which we combine pairs of $\{z(\lambda_t), \gamma_t\}$.

Let us show now that γ may attain $\max\{\gamma_t \mid t \in P\}$. For the sake of simplicity, in the case of binary branching, Theorem 3.2 can be formulated as follows.

If $z^*[S, T-k] - z(\lambda^-) \leq \gamma^- \leq \varepsilon$, and $z^*[S+k, T] - z(\lambda^+) \leq \gamma^+ \leq \varepsilon$, for some $\lambda^-, \lambda^+ \in [0, N]$ and some γ^- and γ^+, then $\max\{z(\lambda^-) + \gamma^-, z(\lambda^+) + \gamma^+\} - \max\{z(\lambda^-), z(\lambda^+)\} = \gamma \leq \max\{\gamma^-, \gamma^+\} \leq \varepsilon$.

Now we can construct an example for which $z^*[S, T] = \max\{z(\lambda^-) + \gamma^-, z(\lambda^+) + \gamma^+\}$, and therefore we can assert that the γ in Theorem 3.2 is the best possible. For example, suppose that $\varepsilon = 12$, $z(\lambda^-) = 15$, $\gamma^- = 8$, $z^*[S, T-k] = 23$, $z(\lambda^+) = 13$, $\gamma^+ = 9$, and $z^*[S+k, T] = 21$. Then, $z^*[S, T-k] - z(\lambda^-) = 23 - 15 = 8 \leq \gamma^- = 8 < \varepsilon$, and $z^*[S+k, T] - z(\lambda^+) = 21 - 13 = 8 < \gamma^+ = 9 < \varepsilon$. Moreover, $z^*[S, T] = 23 = \max\{z(\lambda^-) + \gamma^-, z(\lambda^+) + \gamma^+\} = \max\{15 + 8, 13 + 9\} = 23$ with $\max\{z(\lambda^-), z(\lambda^+)\} = \max\{15, 13\} = 15$, and $\gamma = \max\{15 + 8, 13 + 9\} - \max\{15, 13\} = 8 < \max\{\gamma^-, \gamma^+\} = \max\{8, 9\} = 9$.

For the sake of completeness we prove the "minimization" variant of Theorem 3.2. Let us consider the following problem:

$$\min\{z(\lambda) \mid \lambda \in S\} = z^*(S).$$

Theorem 3.3. *Let z be an arbitrary function defined on a finite set S and let $S = \bigcup_{t \in P} S_t$ with $\min\{z(\lambda) \mid \lambda \in S_t\} = z^*(S_t)$, for $t \in P = \{1, \ldots, p\}$. Then for any $\varepsilon \geq 0$ the following assertion holds.*

If $z(\lambda_t) - z^*(S_t) \leq \gamma_t \leq \varepsilon$ for some $\lambda_t \in S$ and for all $t \in P$, then $\min\{z(\lambda_t) \mid t \in P\} - z^*(S) \leq \min\{z(\lambda_t) \mid t \in P\} - \min\{z(\lambda_t) - \gamma_t \mid t \in P\} = \gamma \leq \max\{\gamma_t \mid t \in P\} \leq \varepsilon$.

Proof. $\min\{z(\lambda_t) \mid t \in P\} - z^*(S) = \min\{z(\lambda_t) \mid t \in P\} - \min\{z^*(S_t) \mid t \in P\} \leq \min\{z(\lambda_t) \mid t \in P\} - \min\{z(\lambda_t) - \gamma_t \mid t \in P\} = \min\{z(\lambda_t) \mid t \in P\} + \max\{\gamma_t + [-z(\lambda_t)] \mid t \in P\} \leq \min\{z(\lambda_t) \mid t \in P\} + \max\{\gamma_t \mid t \in P\} + \max\{[-z(\lambda_t)] \mid t \in P\} = \min\{z(\lambda_t) \mid t \in P\} + \max\{\gamma_t \mid t \in P\} - \min\{[z(\lambda_t)] \mid t \in P\} = \max\{\gamma_t \mid t \in P\}$. $\square$

The main step of the DC algorithm, to be formulated in Sect. 3.2, is called Procedure DC(). The input parameters of Procedure DC() are an interval $[S, T]$, and a prescribed value of ε; the output parameters are λ and γ, with $\lambda \in [0, N]$ and $z^*[S, T] - z(\lambda) \leq \gamma \leq \varepsilon$. The value of γ is an upper bound for the accuracy of $z^*[S, T] - z(\lambda)$, and may sometimes be smaller than the prescribed accuracy ε. The procedure starts with trying to make the interval $[S, T]$ as small as possible by using Corollary 2.3(a) and (b). If this is not possible, the interval is partitioned into two subintervals. Then, with the help of Lemmas 3.1(c) and (d) it may be possible to narrow one of the two subintervals. If this is not possible, the Procedure DC( ) will use the following branching rule.

Branching Rule. For $k \in \arg\min\{\min[\delta^-(S, T, i), \delta^+(S, T, i)] \mid i \in T \backslash S\}$, split the interval $[S, T]$ into two subintervals $[S+k, T], [S, T-k]$, and use the prescribed accuracy ε of $[S, T]$ for both subintervals.

Our choice for the branching variable $k \in T \backslash S$ is motivated by the observation that $\delta^+(S, T, r^+) \leq \delta^+(S, T-k, r^+)$ and $\delta^-(S, T, r^-) \leq \delta^-(S+k, T, r^-)$, following straightforwardly from the submodularity of z. Hence, the values of δ^+, δ^-, for given r^+, r^-, are seen to increase monotonically with successive branchings. Our choice is aimed at making the right-hand sides δ^+, δ^- as large as possible after branching (and if possible nonnegative), with the purpose of increasing the "probability" of satisfying the preservation rules (see Corollary 2.3). Moreover, this branching rule makes the upper bound for the difference between a γ-maximum and a global maximum as small as possible.

Note that in Procedure DC() λ need not be in the interval $[S, T]$. Notice that in most BnB algorithms a solution for a subproblem is searched inside the solution space of that subproblem. From the proofs of Lemma 3.1 and Theorem 3.2 it can be seen that this is not necessary here. For any prescribed accuracy ε the Procedure DC() reads now as follows.

In Sect. 3.3 we will illustrate this algorithm by solving an instance of the SPLP.

3.2   The DC Algorithm (See [66])

The DC algorithm is a BnB type algorithm and is presented as a recursive procedure (Fig. 3.1).

3.2.1   The Data Correcting Algorithm

Input: A submodular function z on $[\emptyset, N]$ and a prescribed accuracy $\varepsilon_0 \geq 0$.
Output: $\lambda \in [\emptyset, N]$ and $\gamma \geq 0$ such that $z^*[\emptyset, N] - z(\lambda) \leq \gamma \leq \varepsilon_0$.
begin call DC$(\emptyset, N, \varepsilon_0; \lambda, \gamma)$
end;

The correctness of the DC algorithm is shown in the following theorem.

Theorem 3.4. *For any submodular function z defined on the interval $[\emptyset, N]$ and for any accuracy $\varepsilon_0 \geq 0$, the DC algorithm constructs an element $\lambda \in [\emptyset, N]$ and an element $\gamma \geq 0$ such that $z^*[\emptyset, N] - z(\lambda) \leq \gamma \leq \varepsilon_0$.*

Proof. We only need to show that each step of the DC algorithm is correct. The correctness of Step 1 follows from the fact that if $S = T$, then the interval $[S, T]$ contains a unique solution and λ satisfies the prescribed accuracy ε_0 (i.e., $z^*[\emptyset, N] - z(\lambda) = z(\lambda) - z(\lambda) = 0 \leq \gamma \leq \varepsilon_0$). The correctness of Steps 2 and 3 follows from Lemma 3.1b and a, respectively; the correctness of Steps 4 and 5 follows from Lemma 3.1d and c, respectively; the correctness of Step 6 follows from Theorem 3.2. So, if the Procedure DC() is called with the arguments $\emptyset$, N, and ε_0, then, when it is finished, $z^*[\emptyset, N] - z(\lambda) \leq \gamma \leq \varepsilon_0$ holds. $\square$

It is possible to make the DC algorithm more efficient if we fathom subproblems by using upper bounds. For subproblems of the form $\min\{z(\lambda) \mid \lambda \in [S, T]\} = z^*[S, T]$, the following lemma, due to [89, 105], provides two upper bounds.

Procedure $DC(S, T, \varepsilon; \lambda, \gamma)$

Input: A submodular function z on the interval $[S, T]$, $\varepsilon \geq 0$.

Output: $\lambda \in [\emptyset, N]$ and $\gamma \geq 0$ such that $z^*[S, T] - z(\lambda) \leq \gamma \leq \varepsilon$.

begin Step 1: if $S = T$

                then begin $\lambda := S; \gamma := 0;$

                         goto Step 7;

              end

      Step 2: Calculate δ^+ and r^+;

           if $\delta^+ \geq 0$

           then begin call $DC(S + r^+, T, \varepsilon; \lambda, \gamma);$

                    {Lemma 3.1b} **goto** Step 7;

           end

      Step 3: Calculate δ^- and r^-;

           if $\delta^- \geq 0$

           then begin call $DC(S, T - r^-, \varepsilon; \lambda, \gamma);$

                    {Lemma 3.1a} **goto** Step 7;

           end

      Step 4: if $\delta^+ \leq \varepsilon$

           then begin call $DC(S + r^+, T, \varepsilon - \delta^+; \lambda, \gamma);$

                    $\gamma := \gamma + \delta^+$ {Lemma 3.1d};

                    goto Step 7;

           end

      Step 5: if $\delta^- \leq \varepsilon$

           then begin call $DC(S, T - r^-, \varepsilon - \delta^-; \lambda, \gamma);$

                    $\gamma := \gamma + \delta^-$ {Lemma 3.1c};

                    goto Step 7;

           end

      Step 6: Select $k \in T \backslash S$ (Branching Rule)

           call $DC(S + k, T, \varepsilon; \lambda^+, \gamma^+)$

           call $DC(S, T - k, \varepsilon; \lambda^-, \gamma^-)$

           $\lambda := \arg\max\{z(\lambda^-), z(\lambda^+)\}$ {Theorem 3.2}

           $\gamma := \max\{z(\lambda^+) + \gamma^+, z(\lambda^-) + \gamma^-\} - \max\{z(\lambda^+), z(\lambda^-)\}$

      Step 7: $\{z^*[S, T] - z(\lambda) \leq \gamma \leq \varepsilon\}$ (with $z(\lambda) = \max\{z(\lambda^+), z(\lambda^-)\}$)

end;

Fig. 3.1 Procedure DC

Lemma 3.2. *If* $z(S) - z(S + i) < 0$ *and* $z(T) - z(T - i) < 0$ *for all* $i \in T \setminus S$, *then*

$$ub_1 = z(S) - \sum_{i \in T \backslash S} [z(S) - z(S + i)] \geq z^*[S, T],$$

and

$$ub_2 = z(S) - \sum_{i \in T \backslash S} [z(T) - z(T - i)] \geq z^*[S, T].$$

Proof. We will prove only ub_1 because the proof of ub_2 is similar. From Theorem 2.1 (iv) we have that $z(T) \le z(S) - \sum_{i \in T \setminus S}[z(S) - z(S+i)]$ for $S \subseteq T \subseteq N$. Let X be a set in $[S, T]$ such that $z(X) = z^*[S, T]$. Then also $z(X) \le z(S) - \sum_{i \in X \setminus S}[z(S) - z(S+i)] \le z(S) - \sum_{i \in T \setminus S}[z(S) - z(S+i)] \ge z^*[S, T]$ since $z(S) - z(S+i) < 0$ for all $i \in T \setminus S$.   □

We next explain how to incorporate such an upper bound into the DC algorithm. During the running of the DC program we keep a global variable β in the subset of N that has the highest function value found so far. Then we can include a Step 3a after Step 3 in Procedure DC().

Step 3a: Calculate $ub := \min\{ub_1, ub_2\}$;
   if $ub - z(\beta) \le \varepsilon$
   then begin $\lambda := \beta; \gamma := ub - z(\beta)$;
       goto Step 7;
   end

We will refer to the upper bound ub defined in Step 3a as the Khachaturov–Minoux bound. It is obvious that λ and γ satisfy $z^*[S, T] - z(\lambda) \le ub - z(\beta) = \gamma \le \varepsilon$. Note that in this case, β is, in general, not an element of the interval $[S, T]$.

The DC algorithm can also be used as a fast greedy heuristic. If the prescribed accuracy ε_0 is very large, branchings never occur at Step 6; the interval $[S, T]$ is halved in every recursive call of the algorithm until $S = T$, and a "greedy" solution is found. Moreover, the calculated accuracy γ gives insight into the quality of the solution obtained, as it is an upper bound for the difference in value of the solution obtained and an optimal solution. Note that, thanks to Steps 2 and 3, the "greedy" solution found by the DC algorithm when a large ε is specified is in general better than the one obtained by a standard or accelerated greedy algorithm like the ones described in [105].

3.3   The SPLP: An Illustration of the DC Algorithm

Recall that the objective function of SPLP is supermodular. The DC algorithm is used for the determination of a global minimum (0-minimum) and a 2-minimum for the SPL problem of which the data are presented in Table 2.1 (see Sect. 2.6).

The recursive solution trees for the cases $\varepsilon_0 = 0$ and 2 are depicted in Figs. 3.2 and 3.3, respectively. Each subproblem is represented by a box in which the values of the input and the output parameters are shown. At each arc of the trees the corresponding steps of the Procedure DC() are indicated. In Fig. 3.2 the prescribed accuracy $\varepsilon_0 = 0$ is not yet satisfied at the second level of the tree, so that a branching is needed. In the case of $\varepsilon_0 = 2$, the DC algorithm applies the branching rule at the third level because after the second level the value of the current accuracy is equal to 1 ($\varepsilon = 1$). An improved version of the DC algorithm applied to the SPL problem is presented in [69] and based on the pseudo-Boolean approach to the SPLP (see Chap. 4 in this book).

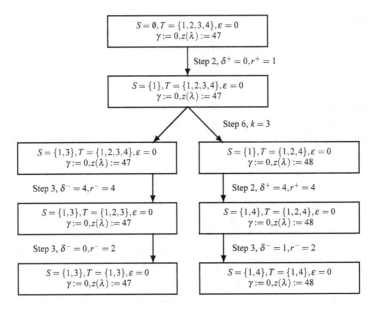

Fig. 3.2 The recursive solution tree for $\varepsilon_0 = 0$

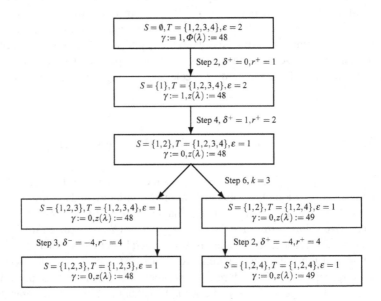

Fig. 3.3 The recursive solution tree for $\varepsilon_0 = 2$

3.4 Computational Experiments with the QCP Problem and Quadratic Zero-One Optimization Problem: A Brief Review

The QCP problem can be described as follows (see Sect. 2.2 and [102]). Given nonnegative real numbers q_{ij} and real numbers p_i with $i, j \in N = \{1, 2, \ldots, n\}$, the QCP is the problem of finding a subset $S \subset N$ such that the function

$$z(S) = \sum_{j \in S} p_i - \frac{1}{2} \sum_{i,j \in S} q_{ij}$$

will be maximized. As a special case we have the max-cut problem (MCP), to be described as follows. Consider an edge weighted undirected graph $U(N, E)$ with edge weights $w_{ij} \geq 0$, $ij \in E$. Define a cut $\delta(T)$ as the edge set containing all the edges with one end in T and the other end in $N \setminus T$. Define further the weight of a cut as the sum of the edge weights in the cut. The MCP is the problem of finding a cut, and thus a partition, with the maximum possible weight. The MCP is a special case of the QCP, namely take

$$p_i = \sum_{i \in V} w_{ij} \text{ and } q_{ij} = 2w_{ij}.$$

An instance of the QCP is defined by an integer positive n, a vector of real numbers P_i, $i = 1, \ldots, n$, and a symmetric matrix $Q = ||q_{ij}||$, $i, j = 1, \ldots, n$ with nonnegative entries.

The QCP and the MCP arise in many real world applications [99, and references within] such as capital budgeting, time tabling, communication scheduling, statistical physics, design of printed circuit boards, and VLSI design (see also [8, 24, 102]). Since the MCP is a special case of the QCP, the QCP is also NP-hard, Karp [87]. An α-approximation algorithm is a polynomial-time algorithm that always finds a feasible solution with an objective function value within a factor α of optimal [135]. The best known α-approximation algorithm for MCP gives $\alpha = 0.87856$ [135]. On the negative side though, Håstad [82] has shown that there can be no 0.941-approximation algorithm for MCP unless $P = NP$. In other words, to solve the MCP with prescribed accuracy within 5.9 % is an NP-hard problem.

The earliest formulation of the QCP (see [79]) in terms of an unconstrained quadratic zero-one programming problem (QZOP) is the following pseudo-Boolean formulation:

$$\max\left(\sum_{i=1}^{n} p_i x_i - \frac{1}{2} \sum_{i=1}^{n} \sum_{j=1}^{n} q_{ij} x_i x_j \mid x \in \{0, 1\}^n\right).$$

Since $x_i^2 = x_i$ we can assume that the diagonal of $Q = ||q_{ij}||$ is zero.

We would like to make an important remark: the equivalence between QZOP and the MCP has been pointed out in [79] (see also [8]). Since the MCP is a special case of the QCP, the QZOP and the QCP are equivalent also. It means that in quadratic time for each instance of the QZOP we can find an instance of the QCP such that they have the same sets of feasible solutions and the same values of densities. We will use this remark as a motivation of the comparability of our computational experiments with the QCP instances and either QZOP or MCP instances reviewed below through the values of densities of the corresponding instances (see [72]).

A mixed-integer programming (MIP) formulation can be found in [113]. In this formulation the quadratic term is replaced by a linear one and a number of linear constraints:

$$\max\left(\sum_{i=1}^{n} p_i x_i - \frac{1}{2} \sum_{i=1}^{n} \sum_{j=1}^{n} q_{ij} y_{ij} \mid x_i + x_j - y_{ij} \leq 1; \right.$$

$$\left. \text{for } i, j = 1, \ldots, n; \ x \in \{0,1\}^n, \ y \in \{0,1\}^{n \times n} \right).$$

Another MIP formulation is given in [102]:

$$\max\left(\sum_{i=1}^{n} p_i x_i - \lambda \mid \lambda \geq \sum_{(i,j) \in E(T) \cup \delta(T)} q_{ij}(x_i + x_j - 1) \right.$$

$$\left. \text{for } T \subseteq N; x \in \{0,1\}^n, \lambda \geq 0 \right)$$

where

$$E(T) = \{(i, j) \mid i \in T, \ j \in T, \ q_{ij} > 0\}$$

and

$$\delta(T) = \{(i, j) \mid i \in T, \ j \in N \backslash T, \ q_{ij} > 0\}.$$

An advantage of the latter formulation over Padberg's formulation [113] is a smaller number of variables, although an exponential number of constraints is required. The exponential number of constraints makes it impossible to solve the full formulation for large instances. In order to overcome this difficulty Lee et al. [102] start with only a small subset of the vertex set constraints $(\lambda \geq \sum_{(i,j) \in E(T) \cup \delta(T)} q_{ij}(x_i + x_j - 1))$ and generate violated ones as soon as they are needed. Therefore they need to solve the problem of recognizing violated constraints, i.e., a separation problem for the vertex set constraints in their branch and cut algorithm. The separation problem is a typical part of branch and cut algorithms based on the polyhedral approach in combinatorial optimization. Boros and Hammer [21] have shown that the corresponding separation problems are polynomially solvable for a wide class of QZOPs.

The methods of computational studies of the QZOP can be classified into the following groups [99]: BnB methods [16, 114], linear programming-based methods (branch and cut algorithms [8, 102]), eigenvalue-based methods, and approaches via semidefinite programming [119, 120]. We will not discuss all of these approaches but restrict ourselves to one important remark. We have not found any computational study of *exact optimal solutions* for the QCP or QZOP for *dense* graphs in which the number of vertices is at least 60. An exception is a specialized exact algorithm for the maximum clique problem (e.g., the stability number problem in the complementary graph) in [25]. They give computational results for problems growing up to 100 variables with any edge density. However, the maximum clique problem is a special case of the QZOP. Billionnet and Sutter [16] gave a comprehensive analysis of computational results published in [8, 24, 25, 86, 114, 134]. For example, Barahona et al. (see Table 3 in [8]) as well as Pardalos and Rodgers (see Table 5.4 in [114]) reported computational results for dense QZOPs with up to 30 vertices; Lee et al. (see Table 1 in [102]) reported computational results for dense QCPs with 40 vertices. Chardaire and Sutter (see Table 1 in [26]) reported computational results for dense QZOPs with up to 50 vertices. For 75 vertices their algorithm only finds the exact optimum for 5 instances out of possible 10. For 100 vertices, they can only find the exact optimum for just one out of ten instances [26]. Moreover, the general conclusion of all published computational studies can be summarized as follows [120]: "When the edge density is decreased, the polyhedral bound is slightly better. On the other hand, increasing the density makes the polyhedral bound poor." In other words, for all the above-mentioned methods, average calculation times grow as edge densities increase. Glover et al. [55] have reported computational experiments with the *adaptive memory tabu search* algorithm for the QZOP on dense graphs with 200 and 500 vertices, and they conclude that these problems are very difficult to solve by current *exact* methods: "Here, however, we have no proof of optimality, since these problems are beyond the scope of those that can be handled within practical time limits by exact algorithms" [55]. In the next section we present computational experiments with the DCA based on the PP algorithm (see [72]). We have found a set of "threshold" QCP instances on dense graphs with up to 400 vertices for which this algorithm solved them to optimality within 10 min on a standard PC. In concluding sections we improve this algorithm and report computational results with "threshold" QCP instances.

3.5 The QCP: Computational Experiments

We have obtained computational results using randomly generated connected graphs with the number of vertices varying from 40 to 80 and edge density d varying from 10 % to 100 % (i.e. $d \in [0.1, 1.0]$). Here the edge density is $d = \frac{|E|}{\binom{m}{2}}$, $|E|$ is the number of random generated edges and $\binom{m}{2}$ is the number of edges in the corresponding simple complete graph. The values of p_i and q_{ij} are uniformly distributed in the intervals [0,100] and [1,100], respectively. The computational results are summarized in Table 3.1. We have tested the DC algorithm on the QCP test problems from [102], and have made a comparison between our results and those from [102].

Table 3.1 The comparison of computational results

Prob.	Time average (s)		# of generated subpr.			# of fathomed subpr.		
	Lee et al.	DC	Minimum	Average	Maximum	Minimum	Average	Maximum
40/2	0.97	0.10	618	797	972	306	396	481
40/3	2.09	0.08	470	640	793	235	313	385
40/4	6.79	0.05	430	539	735	204	258	354
40/5	6.63	0.028	428	497	584	201	231	278
40/6	8.62	0.038	340	387	434	153	173	192
40/7	11.40	0.030	204	216	267	85	100	116
40/8	14.57	0.028	217	261	292	80	95	103
40/9	8.46	0.012	107	154	223	34	42	56
40	13.89	0.004	119	160	213	33	38	48
50/1	0.56	0.29	1,354	1,885	2,525	686	945	1,258
50/2	5.36	0.45	2,100	2,778	3,919	1,042	1,393	1,971
50/3	16.19	0.27	1,671	2,074	2,565	814	1,019	1,268
50/4	95.32	0.18	1,183	1,576	1,976	576	755	950
50/5	38.65	0.08	870	943	1,051	414	447	502
50/6	43.01	0.07	646	725	798	291	321	345
50/7	48.07	0.05	610	648	714	245	270	294
60/2	12.11	1.56	5,470	8,635	11,527	2,718	4,303	5,744
60/3	183.02	0.71	3,481	5,069	7,005	1,736	2,519	3,478
60/4	150.50	0.39	2,450	3,037	3,895	1,221	1,503	1,917
60/5	137.22	0.22	1,701	2,080	2,532	825	1,012	1,236
70/2	437.74	4.89	15,823	23,953	34,998	7,909	11,971	17,486
70/3	367.50	1.91	9,559	11,105	13,968	4,769	5,540	6,967
80/1	20.87	28.12	55,517	92,836	132,447	27,771	46,418	66,228
80/2	864.27	17.10	64,261	66,460	68,372	32,102	33,202	34,160

Each problem set is labeled by the number of vertices of the graph together with their densities d. For example, problem 50/7 refers to graphs with 50 vertices and density $d = 0.7$ (or 70%), problem 40 refers to complete graphs with 40 vertices. For each combination of density and number of vertices, five random problems were solved. The column "Lee et al." of Table 3.1 contains the average computational times for the problems on a RISC 6000 workstation as given in [102]. The DC algorithm was coded by means of Turbo Pascal 6.0 and was executed on a PC with a 133 MHz processor. Cells with "min," "avg," and "max" in Table 3.1 show minimum, maximum, and average performances of two statistics for the DC algorithm: "the number of generated subproblems" solved, and "the number of fathomed subproblems" indicating the number of subproblems discarded by means of the upper bounds ub_1 and ub_2 from Lemma 3.2. When the graph has at most 40 vertices, the problem is very easy, and the calculation times are less than 0.05 s. For sproblems with at least 40 vertices the average calculation times grow exponentially with decreasing values of the density d (see Fig. 3.4) for all values of ε_0. This behavior differs from the results of the algorithm from [102]; their

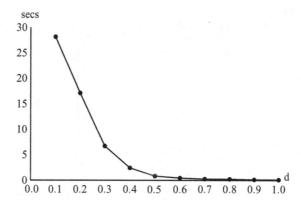

Fig. 3.4 Average calculation time in seconds against the density d (case: $m = 80$, $\varepsilon_0 = 0$)

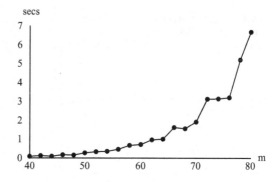

Fig. 3.5 Average calculation time in seconds against the number of vertices m (case: $d = 0.3$, $\varepsilon_0 = 0$)

calculation times grow with increasing densities. For problems with density more than 10 % our algorithm is faster than the algorithm from [102]. For one problem (80/1) with density equal to 10 % our algorithm uses more time.

Some typical properties of the behavior of the DC algorithm are shown in Figs. 3.4–3.6. In Fig. 3.5 it can be seen that the calculation time of the DC algorithm grows exponentially when the number of vertices increases. This is to be expected since general QCPs are NP-hard. Figure 3.6 shows how the calculation times of the DC algorithm depend on the value of ε_0. We have used different prescribed accuracies varying from 0 % to 5 % of the value of the global minimum.

In all experiments with $\varepsilon_0 > 0$ the maximum value of the calculated γ (denoted by $\gamma_{\max}$) is at most $0.01949 \times z^*[\emptyset, N]$, i.e., within 2 % of a global minimum. Moreover, for all test problems with density at least 30 % ($d \geq 0.3$), we obtained $\gamma_{\max} = 0$, that is, we found an exact global minimum with a calculation time of at most 5 s. In Fig. 3.7, $\gamma_{\max}$ is depicted for various values of ε_0.

Prof. Fabio Tardella suggested that in case of the QCP the diagonal dominance of the matrix might have a great influence on the calculation times (see [66]). Assuming that all p_i are positive, the *diagonal dominance* (*dd*) is defined as $dd_i = \frac{p_i}{2\sum_{j \neq i} q_{ij}}, i \in N$, i.e., it is the quotient of the "main diagonal entry" p_i and the sum of

Fig. 3.6 Average time in seconds against prescribed accuracy ε_0 (case: $m = 80, d = 0.2$)

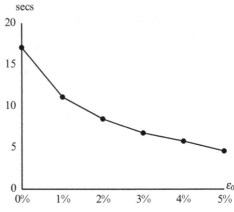

Fig. 3.7 γ_{max} as percentage of the value of a global minimum

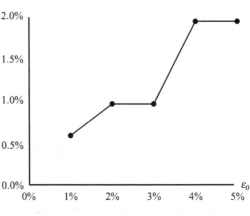

Table 3.2 The distribution of the diagonal dominances

dd	$[0.05, 0.2]$	$[0.2, 0.8]$	$[0.8, 25]$
Percentage	78.8	17.2	4.0

the off diagonal entries in the i-th row and column of $Q = ||q_{ij}||$. We have calculated the diagonal dominance values of the instances from Table 3.1. The results of these calculations are presented in Table 3.2. The first column shows that 78.8 % of the diagonal elements have values in the interval $[0.05, 0.2]$; the meaning of columns 2 and 3 are similar.

We have studied the influence of the diagonal dominance on the average calculation time of the DC algorithm for the following randomly generated instances. The number of vertices m varies from 40 to 80, the edge densities d are chosen in the interval $[0.1, 1.0]$, and the edge weights are randomly generated from the interval $[1, 100]$, just as in Table 3.1. The weights of the vertices p_i, however, are calculated from the edge weights by using a constant dd for all vertices in the same instance, namely $p_i = 2 \text{dd} \sum_{j \neq i} q_{ij}, i \in I^0$. The results for the case $m = 40$ are shown in Fig. 3.8.

Fig. 3.8 Average time in seconds against diagonal dominance (cases: $m = 40, d = 0.3, 0.4, \ldots, 1.0$)

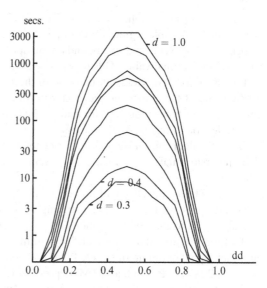

The calculation times grow exponentially with increasing values of the density d, and for fixed d they grow exponentially if dd comes close to 0.5. The maximum calculation time is attained for dd $= 0.5$ and $d = 1.0$. This is the case of a "pure" max-cut problem on a complete graph. Recall that for the instances from [102], as shown in Fig. 3.4, the calculation times decrease by increasing values of the density. Notice that this phenomenon does not occur in case of constant dd.

We may conclude that diagonal dominance is a good "yardstick" for measuring the intractability of instances of the QCP. For example, our randomly generated instances with a constant dd for all vertices from the interval union $[0.05, 0.2] \cup [0.8, 1.0]$ can be classified as "easy" instances of the QCP and from the interval $[0.4, 0.6]$ as "hard" ones. In all our experiments with constant dd, the effect of exponentially increasing calculation times with increasing values of m (see Fig. 3.5), and the exponentially decreasing calculation times with increasing values of ε_0 (see Fig. 3.6) is preserved.

3.6 Remarks for the DC Algorithm

Corollary 2.2 can be considered as the basis of our data correcting algorithm. It states that if an interval $[S, T]$ is split into $[S, T - k]$ and $[S + k, T]$, then the difference between the submodular function values $z(S)$ and $z(S + k)$, or between the values of $z(T)$ and $z(T - k)$ is an upper bound for the difference of the (unknown!) optimal values on the two subintervals. This difference is used for "correcting" the current data (values of z) in the algorithm.

Another keystone in this chapter is Theorem 3.2. For any list of subsets $S_t, t \in P$ that cover the feasible region, for example, $S = \cup_{t \in P} S_t$, it enables us to derive a new upper bound between an upper bound on the optimal value of the objective function on S and the optimal value. This new and sharper upper bound, when implemented in the DC algorithm, yields an increase in the calculated accuracy and a decrease in the value of the associated parameter ε. Moreover, this bound can also be built into other BnB type algorithms for finding the best calculated accuracy based on the accuracies for all resolved subproblems.

The DC algorithm presented in this chapter is a recursive BnB type algorithm. This recursion makes it possible to present a short proof of its correctness; see Theorem 3.4.

We have tested the DC algorithm on cases of the QCP, enabling comparison with the results presented in [102]. The striking computational result is the ability of the DC algorithm to find exact solutions for instances with densities larger than 30 % and with a prescribed accuracy of 5 % within fractions of a second. For example, an exact global optimum of the QCP with 80 vertices and 100 % density was found within 0.22 s on a PC with a 133 MHz processor.

Benati [11] has applied the DC algorithm for solving medium-sized instances of the Uncapacitated Competitive Location Problem with probabilistic customers' behavior.

We point out that when the value of ε_0 is very large, the DC algorithm behaves as a greedy algorithm.

Therefore, the DC algorithm was more efficient for QCP instances defined on dense graphs. However, we did not answer to the following questions.

1. What are the largest (threshold) QCP instances from [102] defined on dense graphs in a reasonable CPU time by the DC algorithm can be solved?
2. Is it possible to increase the threshold numbers of vertices for the QCP instances by any modification of the DC algorithm?

In the remainder of this section we answer the first question and in the next section we answer the second question. Note that most of the published computational experiments with QZOP instances by 10 min CPU per instance are restricted. So, we have kept 10 min CPU as a reasonable time for finding an exact optimal solution to each QCP instance involved in our computational experiments.

We have tested the DC algorithm for QCPs on a Pentium processor running at 300 MHz with 64 MB memory. All algorithms are implemented in Delphi 3.

The threshold QCP instances from [102] which can be solved by the DC algorithm within 10 min and are bounded by 300 vertices are shown in Table 3.3.

Prof. Fabio Tardella has proposed a "measure" of intractability of the QCP in connection with the DC algorithm, namely the so-called diagonal dominance. Our computational experiments with the DC algorithm show that instances of the QCP with diagonal dominances from the intervals union $[0.05, 0.2] \cup [0.8, 25]$ can be classified as "easy" instances, and instances with diagonal dominances from the interval $[0.4, 0.6]$ as "hard" to solve.

Table 3.3 "Threshold" QCP instances solved by the DC algorithm within 10 min

Prob.	Average time (s)	
Dens.	100	200
100	0.831/0.752	64.372/38.351
90	1.027/0.916	78.614/49.926
80	1.784/1.108	217.898/162.455
70	2.498/1.593	631.492/376.629
60	3.509/2.874	1,414.103/895.426
50	9.382/5.931	*/1,937.673
40	17.245/10.327	*/*
30	48.013/22.209	*/*
20	195.82/74.841	*/*
10	446.293/95.122	*/*

Prob.	Average time (s)		
Dens.	300	400	500
100	340.681/229.396	1,894.811/1,162.396	*/*
90	794.037/583.754	3,505.892/1,996.544	*/*
80	2,681.973/1,875.603	*/3,604.715	*/*
70	*/3,165.384	*/*	*/*
60	*/*	*/*	*/*

We would like to remark that the DC algorithm can be used for broad classes of combinatorial optimization problems that are reducible to the maximization of a submodular function. Recently, Krause [96] has incorporated our DC algorithm in the general purpose software MATLAB.

Our computational experiments with the QCP problem show that we can substantially reduce the calculation time for data correcting algorithms [59, 66] by recursive application of our main theorem.

3.7 A Generalization of the DC Algorithm: A Multilevel Search in the Hasse Diagram

In the previous section we have reported computational experiments with DC algorithm applied to QCP problem instances by which we are able to solve the QCP instances on dense graphs up to 300 vertices within 10 min on a standard PC. In this section we improve the above-mentioned DC algorithm for the submodular functions by using instead of two neighboring levels in the Hasse diagram [the so-called Preliminary Preservation Algorithm (PPA) of order zero] more deep search (r levels of the Hasse diagram) procedure (the PPA of order r). We study the behavior of the DC algorithm through the number of search levels of the Hasse diagram for the QCP case. Our computational experiments with the QCP instances from [102] show that the "trade off" level of the Hasse diagram is bounded by $r = 3$

(see [72]). Computational experiments with the improved DCA allow us to solve QCP instances on dense graphs with number of vertices up to 500 within 10 min on a standard PC.

In the next section we determine a generalization of the PPA, called the *PPA of order r* (PPAr).

3.7.1   PPA of Order r

The preservations rules in the PPA "look" only one level deep in the Hasse diagram. The following statements allow us to explore the solution space more than one level deep. This may be useful because we obtain additional possibilities for narrowing the original interval [65].

In this section we use Corollary 2.2 in the following formulation.

Corollary 3.1. *Let z be a submodular function on the interval $[S,T] \subseteq [\emptyset, N]$ and let $k \in T \setminus S$. Then the following assertions hold.*

(a) $z^*[S+k,T] - z^*[S,T-k] \leq z(S+k) - z(S) = d_k^+(S)$.
(b) $z^*[S,T-k] - z^*[S+k,T] \leq z(T-k) - z(T) = d_k^-(T)$.

Theorem 3.5. *Let z be a submodular function on $[S,T] \subseteq [\emptyset, N]$ and let $k \in T \setminus S$. Then the following assertions hold.*

(a) For any $t_0^+(k) \in \arg\max\{d_k^+(S+t) : t \in T \setminus (S+k)\}$,
  $z^*[S+k,T] - \max\{z^*[S,T-k], z(S+k)\} \leq \max\{d_k^+(S+t_0^+(k)), 0\}$.

(b) For any $t_0^-(k) \in \arg\max\{d_k^-(T-t) : t \in T \setminus (S+k)\}$,
  $z^*[S,T-k] - \max\{z^*[S+k,T], z(T-k)\} \leq \max\{d_k^-(T-t_0^-(k)), 0\}$.

Proof. We prove only part (a) since the proof of (b) is similar. Let

$$t_1^+(k) \in \arg\max\{z^*[S+k+t,T] : t \in T \setminus (S+k)\}$$

We may represent the partition of $[S,T]$ by means of its subintervals as follows:

$$[S,T] = S \cup \bigcup_{t \in T \setminus S} [S+t,T].$$

Applying this representation on the interval $[S+k,T]$ we have

$$z^*[S+k,T] = \max\{z(S+k), z^*[S+k+t_1^+(k),T]\}.$$

We distinguish now the following two cases:

Case 1: $z(S+k) \leq z^*[S+k+t_1^+(k),T]$. Then $z^*[S+k,T] = z^*[S+k+t_1^+(k),T]$. For any $k \in T \setminus [S+t_1^+(k)]$ Corollary 3.1(a) on the interval $[S+t_1^+(k),T]$ states that:

$$z^*[S+t_1^+(k)+k,T] - z^*[S+t_1^+(k),T-k] \leq d_k^+(S+t_1^+(k)),$$

i.e., after substituting $z^*[S+k,T]$ instead of $z^*[S+k+t_1^+(k),T]$ this inequality can be written as follows:

$$z^*[S+k,T] - z^*[S+t_1^+(k),T-k] \leq d_k^+(S+t_1^+(k)),$$

and taking into account that $z^*[S+t_1^+(k),T-k] \leq z^*[S,T-k]$ we have

$$z^*[S+k,T] - z^*[S,T-k] \leq d_k^+(S+t_1^+(k)).$$

Adding two maximum operations leads to the following inequality

$$z^*[S+k,T] - \max\{z^*[S,T-k],z(S+k)\} \leq \max\{d_k^+(S+t_1^+(k)),0\}.$$

Finally, $d_k^+(S+t_1^+(k)) \leq d_k^+(S+t_0^+(k))$ since $d_k^+(S+t)$ was maximal for $t_0^+(k)$. This gives the required result.

Case 2: $z(S+k) > z^*[S+k+t_1^+(k),T]$.
Then $z^*[S+k,T] = z(S+k)$. Consider the inequality

$$z(S+k) - \max\{z^*[S,T-k],z(S+k)\} \leq 0.$$

Since $z(S+k) = z^*[S+k,T]$ we have

$$z^*[S+k,T] - \max\{z^*[S,T-k],z(S+k)\} \leq 0.$$

Adding a maximum operation with $d_k^+(S+t_0^+(k))$ gives the required result
$$z^*[S+k,T] - \max\{z^*[S,T-k],z(S+k)\} \leq \max\{d_k^+(S+t_0^+(k)),0\}. \qquad \square$$

Corollary 3.2 (Preservation Rules of Order One). *Let z be a submodular function on $[S,T] \subseteq [0,N]$ and let $k \in T \setminus S$. Then the following assertions hold.*
First Preservation Rule of Order One

(a) *If* $\max\{d_k^+(S+t) : t \in T \setminus (S+k)\} \leq 0$, *then*
   $z^*[S,T] = \max\{z^*[S,T-k],z(S+k)\} \geq z^*[S+k,T].$

Second Preservation Rule of Order One

(b) *If* $\max\{d_k^-(T) : t \in T \setminus (S+k)\} \leq 0$, *then*
   $z^*[S,T] = \max\{z^*[S+k,T],z(T-k)\} \geq z^*[S,T-k].$

In the following theorem we show if the current interval $[S,T]$ cannot be narrowed by preservation rules of order one, then the same interval cannot be narrowed by preservation rules of order zero (see the preservation rules in Corollary 2.3). Moreover, if the interval $[S,T]$ can be narrowed by preservation rules of order zero, then this interval can be narrowed by preservation rules of order one. In this sense we will say that preservation rules of order one are *not weaker* than preservations rules of order zero.

Theorem 3.6. *Preservations rules of order one are not weaker than preservations rules of order zero.*

Proof. We compare only first preservation rules of order one and order zero because the proof for the case of second rules is similar.

Assume that the preservation rule of order one is not applicable, i.e., $\max\{d_k^+(S+t) : t \in T\backslash(S+k)\} = d_k^+(S+t_0) > 0$. The definition of submodularity of z implies $d_k^+(S) \geq d_k^+(S+t_0)$. Hence, $d_k^+(S) > 0$ and the first preservation rule is not applicable. In case when the first preservation rule of order zero is applicable, i.e., $d_k^+(S) \leq 0$, we have $0 \geq d_k^+(S) \geq d_k^+(S+t)$ for all $t \in T\backslash(S+k)$, i.e., $\max\{d_k^+(S+t) : t \in T\backslash(S+k)\} \leq 0$. $\qquad\square$

Note that the computational complexity for rules of order one and order zero is different not only in their time complexities but also in their space complexities because together with the preserved interval either $[S+k,T]$ or $[S,T-k]$ we should preserve exactly one additional value either $z(T-k)$ or $z(S+k)$, respectively. This property is also valid for the preservation rules of order $r \geq 1$.

Instead of one level deep (order one) we may "look" r levels deep (order r) with a view to determine whether we can include or exclude an element. To simplify the presentation of the following theorem, we need some new notations describing certain subsets of the interval $[S,T]$. Let

$$M_r^+[S,T] = \{I \in [S,T] : |I\backslash S| \leq r\},$$
$$M_r^-[S,T] = \{I \in [S,T] : |T\backslash I| \leq r\}.$$

The sets $M_r^+[S,T]$ and $M_r^-[S,T]$ are the collection of subsets that contain in the vicinity on one side of the sets S (the bottom of the corresponding Hasse subdiagram) and T (the top of the corresponding Hasse subdiagram) for r levels deep. Define further the collections of sets

$$N_r^+[S,T] = M_r^+[S,T]\backslash M_{r-1}^+[S,T],$$
$$N_r^-[S,T] = M_r^-[S,T]\backslash M_{r-1}^-[S,T]$$

The sets $N_r^+[S,T]$ and $N_r^-[S,T]$ are the collection of sets which are located on the level r above S and below T in the Hasse diagram, respectively. Let $v_r^+[S,T] = \max\{z(I) : I \in M_r^+[S,T]\}$, $v_r^-[S,T] = \max\{z(I) : I \in M_r^-[S,T]\}$, $w_{rk}^+[S,T] = \max\{d_k^+(I) : I \in N_r^+[S+k,T]\}$ and $w_{rk}^-[S,T] = \max\{d_k^-(I) : I \in N_r^-[S,T-k]\}$.

Theorem 3.7. *Let z be a submodular function on* $[S,T] \subseteq [\emptyset,N]$ *with* $k \in T \setminus S$ *and let r be a positive integer. Then the following assertions hold.*

(a) *If* $|N_r^+[S+k,T]| > 0$, *then*
$$z^*[S+k,T] - \max\{z^*[S,T-k], v_r^+[S,T]\} \le \max\{w_{rk}^+[S,T],0\}.$$

(b) *If* $|N_r^-[S,T-k]| > 0$, *then*
$$z^*[S,T-k] - \max\{z^*[S+k,T], v_r^-[S,T]\} \le \max\{w_{rk}^-[S,T],0\}.$$

Proof. We prove only part (a) since the proof of the part (b) is similar. We may represent the partition of interval $[S,T]$ as follows:

$$[S,T] = M_r^+[S,T] \cup \bigcup_{I \in N_r^+[S,T]} [I,T].$$

Applying this representation on the interval $[S+k,T]$ we have

$$z^*[S+k,T] = \max\{v_r^+[S+k,T], \max\{z^*[I+k,T] : I \in N_r^+[S,T]\}\}.$$

Let $I(k) \in \arg\max\{z^*[I+k,T] : I \in N_r^+[S,T]\}$, and let us consider two cases of the last equality:

Case 1. $z^*[I(k)+k,T] \ge v_r^+[S+k,T]$.
Case 2. $z^*[I(k)+k,T] < v_r^+[S+k,T]$.

In the first case $z^*[S+k,T] = z^*[I(k)+k,T]$. For $I(k) \in N_r^+[S,T]$ Corollary 3.1(a) on the interval $[I(k),T]$ states:

$$z^*[I(k)+k,T] - z^*[I(k),T-k] \le d_k^+(I(k)),$$

i.e. in case 1

$$z^*[S+k,T] - z^*[I(k),T-k] \le d_k^+(I(k)).$$

Note for $[I(k),T-k] \subseteq [S,T-k]$ we have $z^*[S,T-k] \ge z^*[I(k),T-k]$. This leads to the following inequality

$$z^*[S+k,T] - z^*[S,T-k] \le d_k^+(I(k)).$$

Adding two maximum operations gives

$$z^*[S+k,T] - \max\{z^*[S,T-k], v_r^+[S+k,T]\} \le \max\{d_k^+(I(k)),0\}.$$

Since $w_{rk}^+[S,T]$ is the maximum of $d_k^+(I)$ for $I \in N_r^+[S+k,T]$, we have the required result.

In the second case $z^*[S+k,T] = v_r^+[S+k,T]$ the following equality holds:

$$z^*[S+k,T] - v_r^+[S+k,T]\} = 0$$

or

$$z^*[S+k,T] - \max\{z^*[S,T-k], v_r^+[S+k,T]\} \le 0.$$

Adding a maximum operation with $w_{rk}^+[S,T]$ completes the proof of case (a)

$$z^*[S+k,T] - \max\{z^*[S,T-k], v_r^+[S+k,T]\} \le \max\{w_{rk}^+[S,T], 0\}. \qquad \square$$

Corollary 3.3 (Preservation Rules of Order r). *Let z be a submodular function on $[S,T] \subseteq [\emptyset, N]$ and let $k \in T \backslash S$. Then the following assertions hold.*

First Preservation Rule of Order r

(a) *If $w_{rk}^+[S,T] \le 0$, then*
$$z^*[S,T] = \max\{z^*[S,T-k], v_r^+[S+k,T]\} \ge z^*[S+k,T].$$

Second Preservation Rule of Order r

(b) *If $w_{rk}^-[S,T] \le 0$, then*
$$z^*[S,T] = \max\{z^*[S+k,T], v_r^-[S,T-k]\} \ge z^*[S,T-k].$$

Note that the analogue of Theorem 3.6 can be proved for preservation rules of order $r-1$ and r as follows. Preservation rules of order r are not weaker than preservations rules of order $r-1$.

Now we can describe the PPA of order r (PPAr). The PPAr behaves in the same manner as the PPA, i.e., it tries to decrease the original interval $[X,Y]$ in which an optimal solution is located. The difference between the two algorithms lies in the fact that the PPA searches only one level deep in the Hasse diagram, while the PPAr searches r levels deep. The PPAr chooses one element to investigate further from either the top or the bottom of the Hasse diagram. We could investigate all vertices from $T \backslash S$ but this would cost too much time. Therefore we use a heuristic to select the element which we investigate further. The element we choose is an element for which it is likely that one of the preservations rules of order r will succeed in including or excluding this element from an optimal solution. The preservations rules of order one apply if $\max\{d_k^+(S+t) : t \in T \backslash (S+k)\} \le 0$ or $\max\{d_k^-(T-t) : t \in T \backslash (S+k)\} \le 0$. So if we want them to apply then we have to choose an element k so as to minimize the values $d_k^+(S+t)$ and $d_k^-(T-t)$. According to an equivalent definition of a submodular function (see [109]), $d_k^+(S) \ge d_k^+(S+t)$, if we choose $d_k^+(S)$ as small as possible, then $d_k^+(S+t)$ will not be large and hopefully negative for all t, and the first preservation rule of order one is more likely to apply. Also if we take k with the smallest value $d_k^-(T)$ then the second preservation rule of order one is more likely to apply. Our computational study (see Sect. 3.7.4) selects the "best" value of r, and therefore shows the relevance of this choice.

It is clear that if we search deep enough, the PPAr will always find an optimal solution to our problem. We just take $r = |Y \backslash X|$, where $[X,Y]$ is the initial interval, and at each step we will be able to include or exclude an element of the initial interval. However, the number of sets we have to examine in this case is not a polynomial function of r.

Let us define two recursive procedures PPArplus and PPArmin by means of which we can try to include and exclude some elements of the initial interval $[X,Y]$.

Procedure $PPArplus(S, T, k, r, maxd)$
begin Calculate $z(S + k)$;
      If $z(B) < z(S + k)$
      then $B \leftarrow S + k$;
      For all $t \in T \backslash (S + k)$ calculate $d_k^+(S + t)$;
      If $d_k^+(S + t) \leq 0$ or $r = 1$
      then $maxd \leftarrow \max\{maxd, d_k^+(S + t)\}$
      else call $PPArplus(S + t, T, k, r - 1, maxd)$;
end

Procedure $PPArmin(S, T, k, r, maxd)$
begin Calculate $z(T - k)$;
      If $z(B) < z(T - k)$
      then $B \leftarrow T - k$;
      For all $t \in T \backslash (S + k)$ calculate $d_k^-(T - t)$;
      If $d_k^-(T - t) \leq 0$ or $r = 1$
      then $maxd \leftarrow \max\{maxd, d_k^-(T - t)\}$
      else call $PPArplus(S, T - t, k, r - 1, maxd)$;
end

The Preliminary Preservation Algorithm of order r.
Input: A submodular function z on $[X, Y]$ of $[\emptyset, N]$
Output: The subinterval $[S, T]$ and the set B such that
        $z^*[X, Y] = \max\{z^*[S, T], z(B)\}$ and
        $\min\{w_{rk}^+[S, T], w_{rk}^-[S, T]\} > 0$ for all $k \in T \backslash S$
Step 0: $S \leftarrow X, T \leftarrow Y, B \leftarrow \emptyset$;
Step 1: call $PPA(X, Y; S, T)$; **goto** Step 2;
Step 2: $d^+ \leftarrow \max\{d_k^+(S) : k \in T \backslash S\}, d^- \leftarrow \max\{d_k^-(T) : k \in T \backslash S\}$;
        If $d^+ < d^-$ **then goto** Step 3 **else goto** Step 4;
Step 3: $k \leftarrow \arg\max\{d_t^+(S) : t \in T \backslash S\}$;
        call $PPArplus(S, T, k, r, maxd)$;
        If $maxd \leq 0$ **then** $T \leftarrow T - k$, **goto** Step 1.
Step 4: $k \leftarrow \arg\max\{d_t^-(T) : t \in T \backslash S\}$;
        call $PPArmin(S, T, k, r, maxd)$;
        If $maxd \leq 0$ **then** $S \leftarrow S + k$, **goto** Step 1.

Note that the PPAr finds a maximum of the submodular function iff the level r of the Hasse diagram is "deeper or equal" to the level on which a STC is located.

3.7.2   The Data Correcting Algorithm Based on the PPAr

In this section we briefly describe the main idea and the structure of the DC algorithm based on the PPAr and abbreviated to DCA(PPAr). The description of the DCA(PPA0) can be found in Sect. 3.2 (see also [66]). We will point out the main differences between the DCA(PPA) and the DCA(PPAr).

Recall that if a submodular function z is not a PP-function, then the PPA terminates with a subinterval $[S,T]$ of $[\emptyset,N]$ with $S \neq T$ containing a maximum of z without knowing its exact location in $[S,T]$. In this case, the postcondition $\min\{d_i^+(S), d_i^-(T) \mid i \in T\backslash S\} = \delta > 0$ of the PPA is satisfied. The basic idea of the DCA is that if a situation occurs for which this postcondition holds, then the data of the current problem will be corrected in such a way that a corrected function z violates at least one of inequalities $d_k^+(S) = \delta > 0$ or $d_p^-(T) = \delta > 0$ for some $k, p \in T\backslash S$. In that manner the PPA can continue. Moreover, each correction of z is carried out in such a way that the new (corrected) function remains submodular. If the PPA stops again without an optimal solution we apply the correcting rules again and so on until the PPA finds an optimal solution. For $k \in T \backslash S$ Corollary 3.1 gives upper bounds for the values $z^*[S,T] - z^*[S,T-k]$, namely, $d_k^+(S)$, and for $z^*[S,T] - z^*[S+k,T]$, namely, $d_k^-(T)$.

So, if we choose to include an element k in the interval $[S,T]$, then we know that the difference between an optimal solution of the original interval and the new one will be smaller than $d_k^+(S)$. A similar interpretation holds for $d_k^-(T)$. It is clear that after at most n corrections we will find an approximate solution $J \in [\emptyset,N]$ such that $z^*[\emptyset,N] \leq z(J) + \varepsilon$, where $\varepsilon = \sum_{i=1}^n \delta_i$ with δ_i equal to either $d_i^+(S)$ or $d_i^-(T)$.

Before the PPA stops there are a few options. First, if we would like to allow a certain prescribed accuracy, say ε_0, of an approximate solution for the current interval $[S,T]$, then after each correction we must check the inequalities $z^*[X,Y] - z^*[S,T] \leq \varepsilon \leq \varepsilon_0$. If $\varepsilon > \varepsilon_0$ then it is possible to look deeper than one level in the Hasse diagram (see the PPAr) *either* to determine whether or not an element belongs to an optimal solution *or* at least to reduce the current values of $d_i^+(S)$ and $d_i^-(T)$, because $w_{rk}^+[S,T] \geq w_{r+1k}^+[S,T]$ and $w_{rk}^+[S,T] \geq w_{rk}^+[S,T-t]$, or $w_{rk}^-[S,T] \geq w_{r+1k}^-[S,T]$ and $w_{rk}^-[S,T] \geq w_{rk}^-[S-t,T]$. We will explore these possibilities in the DCA(PPAr).

Finally, we can divide the current problem into two subproblems by splitting the corresponding interval into $[S+k,T]$ and $[S,T-k]$ for some chosen k, and apply the PPA on each interval separately. The monotonicity property $d_i^+(S) \geq d_i^+(S+t)$ of a submodular function is the base of the following branching rule (see Sect. 3.1 and [66]). Note that $d_i^+(S) = -\delta^-(S,T,i)$ and $d_i^-(S) = -\delta^+(S,T,i)$.

3.7.3 Branching Rule

For $k \in \arg\max\{\max[d_i^+(S), d_i^-(T)] : i \in T\backslash S\}$, *split the interval $[S,T]$ into two subintervals $[S+k,T]$, $[S,T-k]$, and use the prescribed accuracy ε of $[S,T]$ for both intervals.*

To make the DCA more efficient we incorporate improved upper bounds by which we can discard certain subproblems from further consideration. We may discard a subproblem if some optimal value found so far is larger than the upper bound of the subproblem under investigation because the optimal value of this subproblem will never be larger than the optimal value found so far.

Due to [90], the upper bounds ub_1 and ub_2 from Corollary 3.1 can be tightened. Define the following sets of positive numbers: $d^+(S,T) = \{d_i^+(S) : d_i^+(S) > 0, i \in T\backslash S\}$ and $d^-(S,T) = \{d_i^-(T) : d_i^-(T) > 0, i \in T\backslash S\}$. Define further the ordered arrays: $d^+[i]$ is an i-th largest element of $d^+(S,T)$ and $d^-[i]$ is an i-th largest element of $d^-(S,T)$ both for $i = 1,\ldots,|T\backslash S|$. So, $d^+[1] \geq \cdots \geq d^+[|T\backslash S|]$ and $d^-[1] \geq \cdots \geq d^-[|T\backslash S|]$. Let $z^*[S,T,i] = \max\{z(I) : N_i^+[S,T]\}$ which is the optimal value of $z(I)$. Finally, let us consider two functions which describe the behavior of our upper bounds while we add elements to the set S or delete elements from the set T: $f^+(i) = z(S) + \sum_{j=1}^{i} d^+[j]$ and $f^-(i) = z(T) + \sum_{j=1}^{i} d^-[j]$. Hence, $z^*[S,T,i] \leq \min\{f^+(i), f^-(i)\}$. Since $z^*[S,T] = \max\{z^*[S,T,i] : i = 1,\ldots,|T\backslash S|\}$ we have the following upper bound

$$ub = \max\{\min[f^+(i), f^-(i)] : i = 1,\ldots,|T\backslash S|\} \geq z^*[S,T].$$

Now we will describe the DCA. The DCA starts with a submodular function z on the interval $[\emptyset, N]$ and the prescribed accuracy ε_0. A *list of unsolved subproblems* (LUS) is kept during the course of the DCA. Every time a subproblem is further decomposed into smaller subproblems, one of the subproblems is added to the LUS and to the other one the DCA is applied. After a solution has been found to a subproblem, a new subproblem is taken from the LUS, and so on until the LUS is empty. First, the DCA approximates a subproblem by using the PPA. If this does not result in an optimal solution of that subproblem, it first tries to discard the subproblem by using the upper bound, else the subproblem will be either corrected (if $\varepsilon \leq \varepsilon_0$) or (if $\varepsilon > \varepsilon_0$) split up by means of the branching rule.

Note that the corrections are executed implicitly. A correction allows the PPA to continue at least one step since the correction makes the postcondition of the PPA invalid. For instance, if the PPA stops with an interval $[S,T]$, then after increasing (correction) the value of z on $[S, T - k]$ by $d_k^+(S) > 0$ the DCA may discard the subinterval $[S + k, T]$, because $z^*[S + k, T] - [z^*[S, T - k] + d_k^+(S)] \leq 0$. In fact, instead of correcting the function values of the preserved subinterval, the DCA increases the current value of ε with $d_k^+(S)$. In our example, if the value of the current accuracy of the interval $[S,T]$ is equal to ε, then after discarding the subinterval $[S + k, T]$ its value will be equal to $\varepsilon + d_k^+(S)$. These arguments show that the DCA did not change our submodular function explicitly. On the other hand, let $I \in [S + k, T]$, $J \in [S, T - k]$, then the submodularity of z implies $z(I) + z(J) \geq z(I \cap J) + z(I \cup J)$. Since, $I \cap J \in [S, T - k]$ and $I \cup J \in [S + k, T]$ we have $z(I) + [z(J) + d_k^+(S)] \geq [z(I \cap J) + d_k^+(S)] + z(I \cup J)$. Therefore, by correcting the values of z on a subinterval, the DCA preserves the submodularity of z.

Finally, note that using the PPAr instead of the PPA yields two more possibilities: either by narrowing the current interval or by decreasing the current value of ε.

3.7.4   Computational Experiments for the QCP with the DCA(PPAr)

In [66] we have restricted our computational experiments with the number of vertices up to 80 for the QCP instances, since just such instances in [102] are considered. For these instances from [102] we have shown that the average calculation times grow exponentially when the number of vertices increases and cut down exponentially with the increasing values of the density. For example, an exact global optimum of the QCP with 80 vertices and 100 % density was found within 0.22 s on a PC with a 133 MHz processor but for the QCP with 80 vertices and 10 % density 28.12 s is required on the same PC. Therefore, the DCA(PPA0) was more efficient for QCP instances defined on dense graphs. However, we have not answered the following question yet.

Is it possible to increase the threshold numbers of vertices for the QCP instances by the DCA(PPAr)?

In the remainder of this section we answer to this question. We have tested the DCA(PPAr) for QCPs on a Pentium processor running at 300 MHz with 64 MB memory. All algorithms are implemented in Delphi 3.

The largest part of the calculation time is taken by the calculation of the values of $d_k^+(S)$ and $d_k^-(T)$, since they are calculated rather frequently in the course of the algorithm. In case of the QCP we may calculate, for example, the value of $d_k^+(S)$, by calculating, at the first step, the expressions of

$$z(S+k) = \sum_{i \in S+k} p_i - \frac{1}{2} \sum_{i,j \in S+k} q_{ij}$$

and

$$z(S) = \sum_{i \in S} p_i - \frac{1}{2} \sum_{i,j \in S} q_{ij},$$

and, at the second step,

$$d_k^+(S) = z(S+k) - z(S).$$

However, we can simplify the calculating of $d_k^+(S)$ as follows:

$$d_k^+(S) = z(S+k) - z(S)$$

$$= \sum_{i \in S+k} p_i - \frac{1}{2} \sum_{i,j \in S+k} q_{ij} - \left(\sum_{i \in S} p_i - \frac{1}{2} \sum_{i,j \in S} q_{ij} \right)$$

$$= p_k - \frac{1}{2} \sum_{i,j \in S} q_{ij} + \frac{1}{2} \sum_{i,j \in S} q_{ij} - \frac{1}{2} \sum_{i \in S} q_{ik} - \frac{1}{2} \sum_{j \in S} q_{kj}$$

$$= p_k - \frac{1}{2} \sum_{i \in S} q_{ik} - \frac{1}{2} \sum_{j \in S} q_{kj}.$$

Since $q_{kk} = 0$ and $q_{ij} = q_{ji}$ the last expression can be rewritten as

$$d_k^+(S) = p_k - \sum_{i \in S} q_{ik}.$$

Similarly,

$$d_k^-(T) = \sum_{i \in T} q_{ik} - p_k.$$

Note that values of $d_k^+(S)$ and $d_k^-(T)$ must be calculated for successive sets such that $S, S + t_0, S + t_0 + t_1$ etc., and $T, T - t_0, T - t_0 - t_1$ etc. Hence, we can use the previous value for calculating the next one as follows:

$$d_k^+(S+t) = d_k^+(S) - q_{tk} \text{ and } d_k^-(T-t) = d_k^-(T) - q_{tk}.$$

If we compare the two implementations of the DCA(PPAr), namely with the direct calculation of differences between $d_k^+(S+t)$ and $d_k^-(T-t)$, and with the preliminary simplified expression of $d_k^+(S)$, then computational experiments show that the average computational time is reduced, on average, by a factor of 2.

As problem instances we use randomly generated connected graphs having from 50 to 500 vertices and densities 10–100 % which are "statistically" the same as from [102]. The *density* is defined as

$$d = \frac{|E|}{n(n-1)/2} \cdot 100\%,$$

where $|E|$ is the number of generated edges and $n(n-1)/2$ is the number of edges in a complete simple graph. The data p_i and q_{ij} are uniformly distributed with $p_i \in [0, 100]$ and $q_{ij} \in [1, 100]$. So, we may compare our computational results (see also [66]) to those obtained by Lee et al. [102].

First of all we look at calculation times of the DCA(PPAr) needed for problems varying from 50 to 100 vertices. Since the DCA(PPAr) finds easily an optimal solution to the instances for which an optimal solution as close as possible either to the top or to the bottom of the Hasse diagram is located, we use the *distance*

$$\text{dist}(|I|, n/2) = \frac{||I| - n/2|}{n/2} \cdot 100\%$$

between the calculated optimal solution I and the level $n/2$ of the "main diagonal" of the Hasse diagram in percentages as one of parameters of the "hardness" of our instances by solving them to optimality by the DCA(PPAr).

Intuitively, it is clear that the DCA(PPAr) applied to instances with distances close to 0 % requires more calculation time than the DCA(PPAr) applied to instances with distances close to 100 %. Empirically we have found (see [66]) that for sparse instances from [102] the distance is close to the "main diagonal" of the Hasse diagram (see Fig. 3.9). For sparse instances with densities bounded by 20 % the DCA(PPA0) outperforms the branch-and-cut algorithm from [102], often with speeds 10 times faster and for nonsparse instances with density more than 40 % with speeds 100 times faster. Figure 3.9 shows that the distance grows when the density of instances increases. Therefore, we can expect a decrease in the average calculation time [66] when the density of instances increases (see Fig. 3.10). Figure 3.10 shows that the natural logarithm of the average calculation time is approximately linear.

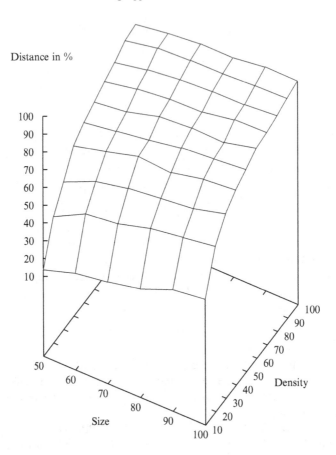

Fig. 3.9 dist($|I|, n/2$) for instances of the QCP with 50–100 vertices and densities 10–100 %

Hence, it is plausible that the average calculation time grows exponentially when the size of instances increases. Moreover, this increasing is more rapid for sparse instances than for dense ones. The threshold QCP instances from [102] which can be solved by the DCA(PPA0) within 10 min in Table 3.3 are shown and bounded by 300 vertices.

We also study the impact of the number r of levels of the PPAr on the average calculation time of the DCA(PPAr). Figure 3.11 shows that searching one or more levels deep does not decrease the average calculation time for non-dense instances ($d < 1.0$). The smallest average calculation time is achieved at level 3 for instances of complete graphs ($d = 1.0$). This fact is explained for all cases by the number of generated subproblems for different levels r (see Fig. 3.12). In Fig. 3.12 it can be seen that in all cases the number of subproblems is decreased when we search deeper, but the decrease percentage of the number of subproblems for levels 0 through 5 is only 14 % for instances with density of 70 % while it is 91 % for instances with density 100 %. Therefore the profit of decreasing the number of

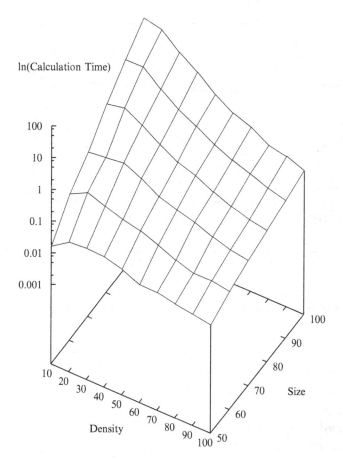

Fig. 3.10 Natural logarithm of the average calculation time (in seconds) for instances of the QCP with 50–100 vertices and densities 10–100 %

subproblems is spent on the additional costs of the average calculation time for the PPAr. More exactly, for dense graphs the balance is positive for search levels 3 and 4. This effect holds also for larger instances (see Fig. 3.13).

In the second part of experiments we consider instances of the QCP with sizes between 100 and 500 vertices and densities between 10 and 100 % which can be solved with a prescribed accuracy of up to 5 % within approximately 10 min. Table 3.4 gives calculation times in seconds for exact/approximate solutions (0 %/5 %). The entries in this table with * could not be solved within 10 min. All instances with sizes above 300 and densities below 50 % could not be solved within 10 min and are not shown in Table 3.4. In all experiments of the second part, the effect of exponentially increasing calculation times with increasing of sizes and decreasing of densities is preserved. Therefore instances of the QCP with 500 vertices and densities between 90 and 100 % are the largest instances which can be solved by the DCA(PPA3) within 10 min on a standard PC.

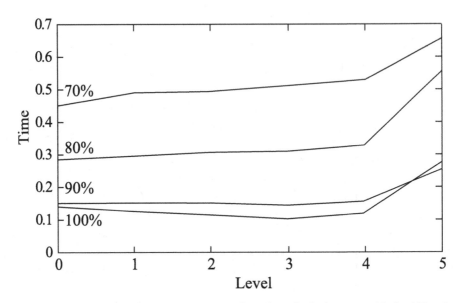

Fig. 3.11 Average calculation times (in seconds) for values of r for instances with size 100 and densities 70–100 %

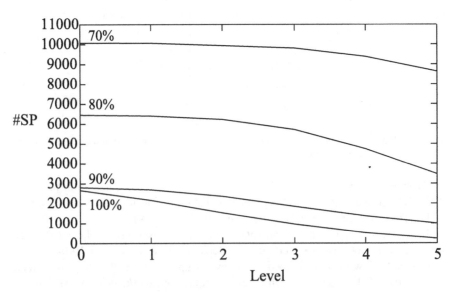

Fig. 3.12 The number of generated subproblems against the level r for instances of the QCP with 100 vertices and densities 70–100 %

The impact of the "diagonal dominance" notion for instances of the QCP is the same as in our previous experiments (see [66] and Sect. 3.5).

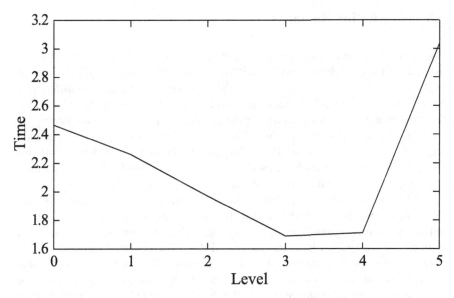

Fig. 3.13 The average calculation time (in seconds) against the level r for instances of the QCP with 200 vertices and density 100 %

Table 3.4 Average calculation times (in seconds) against prescribed accuracies of 0 and 5 % for instances of the QCP with 100–500 vertices and densities 10–100 % within 10 min

Prob. Dens.	Time average (s) 100	200
100	0.098/0.094	2.630/2.444
90	0.138/0.118	3.824/3.607
80	0.280/0.228	9.506/8.186
70	0.393/0.304	17.643/15.693
60	0.731/0.517	86.330/72.931
50	1.752/1.298	345.723/267.445
40	3.457/2.179	*/*
30	11.032/5.880	*/*
20	47.162/17.477	*/*
10	70.081/12.196	*/*

Prob. Dens.	Time average (s) 300	400	500
100	18.316/ 17.179	85.827/ 85.096	229.408/222.883
90	37.931/ 34.972	173.063/166.996	624.925/608.755
80	98.690/ 89.685	679.914/580.789	*/*
70	413.585/364.480	*/*	*/*
60	*/*	*/*	*/*

3.8 Concluding Remarks

Theorem 3.5(a) can be considered as a generalization of Corollary 3.1(a) which is the basis of the DCA(PPAr) and DCA(PPA), respectively. It states that if an interval $[S, T]$ is split into $[S, T - k]$ and $[S + k, T]$, then the maximum value of all differences between the submodular function values z on levels $r + 1$ and r is an *upper bound* for the difference between the unknown optimal value on the discarded subinterval and the maximum of the unknown value of the preserved subinterval and the maximum value of $z(I)$ on r levels of the Hasse diagram. Theorem 3.5(b) can be expounded similarly. These upper bounds are used for implicit "correcting" of the current values of z. In fact, we correct the value of the current accuracy (see [72]).

We have tested the DCA(PPAr) on the QCP instances which "statistically" are the same as in [102]. It is shown that the distance of an optimal solution to an antichain ("main diagonal") of the Hasse diagram is a good measure of an instance difficulty of the QCP at least for the DCA(PPAr). This distance increases a little slower than a linear function against the increasing values of the density for any fixed number of vertices (size). The instances with distances from 0 to 20 % can be categorized as "hard," distances from 30 to 60 % as "difficult," and distances from 70 to 100 % as "easy." In all tested instances the average calculation time grows exponentially for decreasing density values for all prescribed accuracies values. This behavior differs from the results of the branch and cut algorithm in [102]. Their calculation times grow when densities increase. This effect is also indicated for all algorithms based on linear programming (see, e.g., [8, 114, 120]). Our experiments with different levels r of the PPAr show that for the QCP instances from [102], the best level is 3. In addition, this effect will be more evident when the density of the corresponding instances will be as close as possible to 100 %. Note that the largest QCP instances solved by the DCA(PPA0) based on the two neighboring levels in the Hasse diagram within 10 min on a standard PC by 300 vertices are bounded.

Glover et al. [55] reported their computational experiments for binary quadratic programs with *adaptive memory tabu search* procedures. They assumed that the so-called "c" problems with $n = 200$ and 500 "(which are 'statistically' equivalent to the Lee et al. [102] instances defined on dense graphs) to be the most challenging problems reported in the literature to date—far beyond the capabilities of current exact methods and challenging as well for heuristic approaches."

Recently Billionne and Elloumi [17] and Rendl et al. [124] have reported the so called "progress" with solving Max-Cut problem instances on dense graphs with up to 100 and 250 vertices, respectively. This is negligible "achievement" compared to the DCA (PPA3) ability to solve similar problems with up to 500 vertices. It seems that the authors of both papers [17, 124] are overlooked the computational results produced by the data correcting approach to the Quadratic Cost Partition and Max-Cut problems published in [66, 72].

The DCA(PPA3) have solved instances of the QCP up to 500 vertices on dense graphs within 10 min on a standard PC. Since the data correcting approach is applicable for solving the large QCP instances defined on the dense graphs, it will be interesting to investigate a "composition" of data correcting approach and branch-and-cut type algorithms based on mixed integer linear programming for solving large instances of the QCP up for all range of densities.

Chapter 4
Data Correcting Approach for the Simple Plant Location Problem

In this chapter we improve the DC algorithm for general supermodular functions by using a pseudo-Boolean representation of the simple plant location problem (SPLP) presented in the previous chapters. It is common knowledge that exact algorithms for $\mathcal{NP}$-hard problems in general, and for the SPLP in particular, spend only about 10 % of the execution time to find an optimal solution. The remaining time is spent proving the optimality of the solution. In this chapter, our aim is to reduce the amount of time spent proving the optimality of the solution obtained. We propose a data correcting algorithm for the SPLP that is designed to output solutions with a prespecified *acceptable accuracy* ε (see [53]). This means that the difference between the cost of the solution created by the algorithm is at most ε more than the cost of an optimal solution. (Note that $\varepsilon = 0$ results in an exact algorithm for the SPLP, while $\varepsilon = \infty$ results in a fast greedy algorithm.) The objective function of the SPLP is supermodular (see [35]), and so the data correcting algorithm described in Sect. 3.2 (see also [66]) can be used to solve the SPLP. In fact, Sect. 3.3 contains an example to that effect. However, it can be made much more efficient; for example, by using SPLP-specific bounds (used in [44]) and preprocessing rules (used in [3, 46, 89, 91, 118]). The algorithm described here uses a pseudo-Boolean representation of the SPLP, due originally to [80] (see Sect. 4.2). It uses a new *Reduction Procedure* (RP) based on data correcting, which is stronger than the preprocessing rules used in [91] to reduce the original instance to a smaller "core" instance, and then solves it using a procedure based on PP and DC (see Sect. 3.2 and [66]) algorithms. Recently, the RP procedure has been successfully applied to the p-Median Problem (see [74, 75]). Since the new Reduction Procedure is based on a lower bound to the SPLP, we have compared the computational efficiency of this procedure on two different lower bounds. The first lower bound is the well-known Khachaturov–Minoux bound (see Lemma 3.2) which is valid for a general submodular (supermodular) function. The second lower bound is due to [44] which is based on a pair of primal and dual mathematical programming formulations of the SPLP. We show how the use of preprocessing and bounds specific to the SPLP enhance the performance of the data correcting algorithm. This algorithm is based

B. Goldengorin and P.M. Pardalos, *Data Correcting Approaches in Combinatorial Optimization*, SpringerBriefs in Optimization, DOI 10.1007/978-1-4614-5286-7_4, © Boris Goldengorin, Panos M. Pardalos 2012

on two concepts presented in the previous chapter, namely, data correcting and the preliminary preservation procedure (see Chap. 2). Computational experiments with the data correcting algorithm on benchmark instances of the SPLP are also described in this chapter.

4.1 Introduction

Given sets $I = \{1, 2, \ldots, m\}$ of sites in which plants can be located, $J = \{1, 2, \ldots, n\}$ of clients, a vector $F = (f_i)$ of fixed costs for setting up plants at sites $i \in I$, a matrix $C = [c_{ij}]$ of transportation costs from $i \in I$ to $j \in J$, and a unit demand at each client site, the simple plant location problem (SPLP) is the problem of finding a set S, $\emptyset \subset S \subseteq I$, at which plants can be located so that the total cost of satisfying all client demands is minimal. The costs involved in meeting the client demands include the fixed costs of setting up plants, and the transportation cost of supplying a given client from an open plant. We will assume that the capacity at each plant is sufficient to meet the demand of all clients. We will further assume that each client has a demand of one unit, which must be met by one of the opened plants. If a client's demand is different from one unit, we can scale the demand to a unit by scaling the transportation costs accordingly.

A detailed introduction to this problem appears in [35]. SPLP forms the underlying model in several combinatorial problems, like set covering, set partitioning, information retrieval, simplification of logical Boolean expressions, airline crew scheduling, vehicle despatching [32], assortment [14, 56, 64, 84, 116, 117, 128] and is a subproblem for various location analysis problems [122, 123].

The SPLP is NP-hard [35], and many exact and heuristic algorithms to solve the problem have been discussed in the literature. Most of the exact algorithms are based on a mathematical programming formulation of the SPLP. Direct approaches [106, 126] use a BnB approach and the strong linear programming relaxation (SPLR) for computing bounds. However such approaches cannot always output an optimal solution to average-sized SPLP instances in reasonable time. More efficient solution approaches to the SPLP are based on Lagrangian duality [14, 83]. Computational experience of solving the Lagrangian dual using subgradient optimization have been reported in [34, 36], and using Dantzig–Wolfe decomposition in [49]. Computer codes for solving medium-sized SPLP using a mixed-integer programming system are also available [104, 132]. Polyhedral results for the SPLP polytope have been reported in [5, 30, 31, 33, 95, 107, 129]. In theory, these results allow us to solve the SPLP by applying the simplex algorithm to SLPR, with the additional stipulation that a pivot to a new extreme point is allowed only when this new extreme point is integral. Although some computational experience using this method has been reported in the literature [77], efficient implementations of this pivot rule are not available. Results of computational experiments with Lagrangian heuristics for medium-sized instances of the SPLP have been reported in [9]. Large-sized SPLP instances can be solved using algorithms based on refinements to a dual-ascent heuristic procedure to solve the dual of LP-relaxation of the SPLP [93], combined with the use of the complementary slackness conditions to construct primal solutions [44]. An annotated bibliography is available in [98, 122, 123].

It is easy to see that any instance of the SPLP has an optimal solution in which each customer is satisfied by exactly one plant. We believe that Hammer (see [80]) first used this fact to derive a pseudo-Boolean representation of this problem. The pseudo-Boolean function developed in that work has terms that contain both a literal and its complement which can be replaced by a different pseudo-Boolean form containing terms with either only literals or only their complements. This fact is clearly illustrated by an example in [80] and has been generalized in [13]. We find this form easier to manipulate, and hence use Hammer–Beresnev's formulation in this chapter.

The Hammer–Beresnev's pseudo-Boolean representation of the SPLP facilitates the construction of rules to reduce the size of SPLP instances [14, 35, 38, 69, 91, 133, and the references within]. These rules have been used in [67, 68] not only for preprocessing, but also as a tool to either solve a subproblem or reduce its size. They also use information from Hammer–Beresnev's pseudo-Boolean representation of the SPLP to compute efficient branching functions (see [67, 68]). For the sake of simplicity, we use a common depth first BnB scheme in our implementations and Khachaturov–Minoux (see [89, 105]) bound, but the concepts developed herein can easily be implemented in any of the algorithms cited above.

The remainder of this chapter is organized as follows. In Sect. 4.2 we describe Hammer–Beresnev's pseudo-Boolean approach to the SPLP, and use this approach to present Cherenin's Preprocessing Rules (Sect. 4.3). We then describe the ingredients of Data Correcting Approach to the SPLP (Sect. 4.4) and analyze this approach by extensive computational study (Sect. 4.5). We finally summarize this chapter in Sect. 4.6 with concluding remarks.

4.2 A Pseudo-Boolean Approach to SPLP

An instance of the SPLP is described by an m-vector $F = (f_i)$, and a $m \times n$ matrix $C = [c_{ij}]$. We assume that F and C are finite, i.e., $F \in \Re^m$, and $C \in \Re^{mn}$. We will use the $m \times (n+1)$ *augmented matrix* $[F|C]$ as a shorthand for describing an instance of the SPLP. The total cost $f_{[F|C]}(S)$ associated with a solution S consists of two components, the fixed costs $\sum_{i \in S} f_i$, and the transportation costs $\sum_{j \in J} \min\{c_{ij} | i \in S\}$, i.e.,

$$f_{[F|C]}(S) = \sum_{i \in S} f_i + \sum_{j \in J} \min\{c_{ij} | i \in S\},$$

and the SPLP is the problem of finding

$$S^\star \in \arg\min\{f_{[F|C]}(S) : \emptyset \subset S \subseteq I\}. \tag{4.1}$$

An $m \times n$ *ordering matrix* $\Pi = [\pi_{ij}]$ is a matrix each of whose columns $\Pi_j = (\pi_{1j}, \ldots, \pi_{mj})^T$ defines a permutation of $1, \ldots, m$. Given a transportation matrix C, the set of all ordering matrices Π such that $c_{\pi_{1j}j} \leq c_{\pi_{2j}j} \leq \cdots \leq c_{\pi_{mj}j}$, for $j = 1, \ldots, n$, is denoted by perm(C). A vector $(\pi_{1j}, \ldots, \pi_{kj})^T$, $1 \leq k \leq m$, is called a *sub-permutation* of the permutation Π_j.

Defining

$$y_i = \begin{cases} 0 & \text{if } i \in S \\ 1 & \text{otherwise,} \end{cases} \quad \text{for each } i = 1, \ldots, m \tag{4.2}$$

we can indicate any solution S by a vector $\mathbf{y} = (y_1, y_2, \ldots, y_m)$. The fixed cost component of the total cost can be written as

$$\mathscr{F}_F(\mathbf{y}) = \sum_{i=1}^{m} f_i(1 - y_i). \tag{4.3}$$

Given a transportation cost matrix C, and an ordering matrix $\Pi \in$ perm(C), we can denote differences between the transportation costs for each $j \in J$ as

$$\Delta c[0, j] = c_{\pi_{1j}j}, \quad \text{and}$$

$$\Delta c[l, j] = c_{\pi_{(l+1)j}j} - c_{\pi_{lj}j}, \quad l = 1, \ldots, m-1.$$

Then, for each $j \in J$,

$$\min\{c_{ij} | i \in S\} = \Delta c[0, j] + \Delta c[1, j] \cdot y_{\pi_{1j}} + \Delta c[2, j] \cdot y_{\pi_{1j}} \cdot y_{\pi_{2j}}$$

$$+ \cdots + \Delta c[m-1, j] \cdot y_{\pi_{1j}} \cdots y_{\pi_{(m-1)j}}$$

$$= \Delta c[0, j] + \sum_{k=1}^{m-1} \Delta c[k, j] \cdot \prod_{r=1}^{k} y_{\pi_{rj}},$$

so that the transportation cost component of the cost of a solution $\mathbf{y}$ corresponding to an ordering matrix $\Pi \in$ perm(C) is

$$\mathscr{T}_{C,\Pi}(\mathbf{y}) = \sum_{j=1}^{n} \left\{ \Delta c[0, j] + \sum_{k=1}^{m-1} \Delta c[k, j] \cdot \prod_{r=1}^{k} y_{\pi_{rj}} \right\}. \tag{4.4}$$

Lemma 4.1. $\mathscr{T}_{C,\Pi}(\cdot)$ *is identical for all* $\Pi \in$ perm(C).

Proof. Let $\Pi = [\pi_{ij}], \Psi = [\psi_{ij}] \in$ perm(C), and any $\mathbf{y} \in \{0, 1\}^m$. It is sufficient to prove that $\mathscr{T}_{C,\Pi}(\mathbf{y}) = \mathscr{T}_{C,\Psi}(\mathbf{y})$ when

$$\pi_{kl} = \psi_{(k+1)l}, \tag{4.5}$$

$$\pi_{(k+1)l} = \psi_{kl}, \tag{4.6}$$

$$\pi_{ij} = \psi_{ij} \quad \text{if } (i, j) \neq (k, l). \tag{4.7}$$

Then

$$\mathcal{T}_{C,\Pi}(\mathbf{y}) - \mathcal{T}_{C,\Psi}(\mathbf{y}) = (c_{\pi_{(k+1)l}l} - c_{\pi_{kl}l}) \cdot \prod_{i=1}^{k} y_{\pi_{il}} - (c_{\psi_{(k+1)l}l} - c_{\psi_{kl}l}) \cdot \prod_{i=1}^{k} y_{\psi_{il}}.$$

But (4.5) and (4.6) imply that $c_{\pi_{(k+1)l}l} = c_{\pi_{kl}l}$ and $c_{\psi_{(k+1)l}l} = c_{\psi_{kl}l}$ which in turn imply that $\mathcal{T}_{C,\Pi}(\mathbf{y}) = \mathcal{T}_{C,\Psi}(\mathbf{y})$.   □

Combining (4.3) and (4.4), the total cost of a solution $\mathbf{y}$ to the instance $[F|C]$ corresponding to an ordering matrix $\Pi \in \text{perm}(C)$ is

$$f_{[F|C],\Pi}(\mathbf{y}) = \mathcal{F}_F(\mathbf{y}) + \mathcal{T}_{C,\Pi}(\mathbf{y})$$

$$= \sum_{i=1}^{m} f_i(1 - y_i) + \sum_{j=1}^{n} \left\{ \Delta c[0,j] + \sum_{k=1}^{m-1} \Delta c[k,j] \cdot \prod_{r=1}^{k} y_{\pi_{rj}} \right\}. \quad (4.8)$$

Lemma 4.2. *The total cost function $f_{[F|C],\Pi}(\cdot)$ is identical for all $\Pi \in \text{perm}(C)$.*

Proof. This is a direct consequence of Lemma 4.1.   □

A *pseudo-Boolean polynomial* of degree m is a polynomial of the form

$$P(\mathbf{y}) = \sum_{T \in 2^m} \alpha_T \cdot \prod_{i \in T} y_i,$$

where 2^m is the power set of $\{1, 2, \ldots, m\}$, a constant α_T can assume arbitrary values, and y_i is a Boolean variable. We call a pseudo-Boolean polynomial $P(\mathbf{y})$ a *Hammer–Beresnev function* if there exists an SPLP instance $[F|C]$ and $\Pi \in \text{perm}(C)$ such that $P(\mathbf{y}) = f_{[F|C],\Pi}(\mathbf{y})$ for $\mathbf{y} \in \{0,1\}^m$. We denote a Hammer–Beresnev function corresponding to a given SPLP instance $[F|C]$ by $\mathcal{H}_{[F|C]}(\mathbf{y})$ and define it as

$$\mathcal{H}_{[F|C]}(\mathbf{y}) = f_{[F|C],\Pi}(\mathbf{y}) \text{ where } \Pi \in \text{perm}(C). \quad (4.9)$$

Theorem 4.1. *A general pseudo-Boolean function is a Hammer–Beresnev function if and only if*

(a) *All coefficients of the pseudo-Boolean function except those of the linear terms are nonnegative.*
(b) *The sum of the constant term and the coefficients of all the negative linear terms in the pseudo-Boolean function is nonnegative.*

Proof. The "if" statement is trivial. In order to prove the "only if" statement, consider an SPLP instance $[F|C]$, an ordering matrix $\Pi \in \text{perm}(C)$, and a Hammer–Beresnev function $\mathcal{H}_{[F|C]}(\mathbf{y})$ in which there is a nonlinear term of degree k with a negative coefficient. Since nonlinear terms are contributed by the transportation costs only, a nonlinear term with a negative coefficient implies that $\Delta C[k,j]$ for some $j \in \{1, \ldots, n\}$ is negative. But this contradicts the fact that $\Pi \in \text{perm}(C)$.

Next suppose that in $\mathscr{H}_{[F|C]}(\mathbf{y})$, the sum of the constant term and the coefficients of the negative linear terms are negative. This implies that the coefficient of some linear term in the transportation cost function is negative. But this also contradicts the fact that $\Pi \in \mathrm{perm}(C)$. The logic above holds true for all members of $\mathrm{perm}(C)$ as a consequence of Lemma 4.1.   □

Therefore we have shown [1] that the total cost function $f_{[F|C],\Pi}(\cdot)$ is identical for all $\Pi \in \mathrm{perm}(C)$. In other words

$$\mathscr{H}_{[F|C]}(\mathbf{y}) = f_{[F|C],\Pi}(\mathbf{y}) \text{ where } \Pi \in \mathrm{perm}(C). \tag{4.10}$$

We can formulate (4.1) in terms of Hammer–Beresnev functions as

$$\mathbf{y}^* \in \arg\min\{\mathscr{H}_{[F|C]}(\mathbf{y}) : \mathbf{y} \in \{0,1\}^m, \mathbf{y} \neq \mathbf{1}\}. \tag{4.11}$$

Hammer–Beresnev functions assume a key role in the development of the algorithms described in the next sections.

Example 4.1. Consider the SPLP instance:

$$[F|C] = \begin{bmatrix} 7 & 7 & 15 & 10 & 7 & 10 \\ 3 & 10 & 17 & 4 & 11 & 22 \\ 3 & 16 & 7 & 6 & 18 & 14 \\ 6 & 11 & 7 & 6 & 12 & 8 \end{bmatrix}. \tag{4.12}$$

Two of the four possible ordering matrices corresponding to C are

$$\Pi_1 = \begin{bmatrix} 1 & 3 & 2 & 1 & 4 \\ 2 & 4 & 3 & 2 & 1 \\ 4 & 1 & 4 & 4 & 3 \\ 3 & 2 & 1 & 3 & 2 \end{bmatrix} \text{ and } \Pi_2 = \begin{bmatrix} 1 & 4 & 2 & 1 & 4 \\ 2 & 3 & 4 & 2 & 1 \\ 4 & 1 & 3 & 4 & 3 \\ 3 & 2 & 1 & 3 & 2 \end{bmatrix}. \tag{4.13}$$

The Hammer–Beresnev function is $\mathscr{H}_{[F|C]}(\mathbf{y}) = [7(1-y_1) + 3(1-y_2) + 3(1-y_3) + 6(1-y_4)] + [7 + 3y_1 + 1y_1y_2 + 5y_1y_2y_4] + [7 + 0y_3 + 8y_3y_4 + 2y_1y_3y_4] + [4 + 2y_2 + 0y_2y_3 + 4y_2y_3y_4] + [7 + 4y_1 + 1y_1y_2 + 6y_1y_2y_4] + [8 + 2y_4 + 4y_1y_4 + 8y_1y_3y_4]$
$= 52 - y_2 - 3y_3 - 4y_4 + 2y_1y_2 + 8y_3y_4 + 4y_1y_4 + 11y_1y_2y_4 + 10y_1y_3y_4 + 4y_2y_3y_4.$

4.3   Cherenin's Preprocessing Rules

Suppose that the given instance is not recognized to correspond to a known polynomially solvable special case. Then we have to use an exact algorithm for solving this instance. The execution times of exact algorithms for the SPLP are exponential in the parameter m. So any preprocessing rules, i.e., quick methods of reducing the size of the given instance, are of much practical importance.

There are two preprocessing rules available in the literature. The first one, due to Beresnev [13], Cornuejols et al. [35], Dearing et al. [38], and Veselovsky [133] states that if there are two clients that have the same sub-permutations of the transportation costs in any ordering matrix, then they can be aggregated into a single virtual client. The second rule, due to Cornuejols et al. [35] and Dearing et al. [38], states that if the coefficient a_k of the linear term involving y_k is nonnegative, then $y_k = 0$ in an optimal solution, i.e., there exists an optimal solution in which a plant will be opened in that site.

The existing first rule is automatically applied and generalized when we construct a Hammer–Beresnev function. In instance (4.12) we can aggregate clients 1 and 4 by using the first rule. In the following example we see that the same rule can be used to further decrease the number of clients to three.

Example 4.2. Consider the SPLP instance described in (4.12). Combining the first and the fourth client we get the equivalent instance

$$[S|D] = \begin{bmatrix} 7 & 14 & 15 & 10 & 10 \\ 3 & 21 & 17 & 4 & 22 \\ 3 & 34 & 7 & 6 & 14 \\ 6 & 23 & 7 & 6 & 8 \end{bmatrix}.$$

The Hammer–Beresnev function for both these instances is $\mathcal{H}_{[F|C]}(\mathbf{y}) = 52 - y_2 - 3y_3 - 4y_4 + 2y_1y_2 + 8y_3y_4 + 4y_1y_4 + 11y_1y_2y_4 + 10y_1y_3y_4 + 4y_2y_3y_4$. Since this Hammer–Beresnev function can be represented as $\mathcal{H}_{[F|C]}(\mathbf{y}) = [1(1 - y_2) + 3(1 - y_3) + 4(1 - y_4)] + [44 + 0y_1 + 2y_1y_2 + 11y_1y_2y_4] + [0 + 0y_3 + 8y_3y_4 + 4y_2y_3y_4] + [0 + 0y_4 + 4y_1y_4 + 10y_1y_3y_4]$.
the following equivalent instance with *three* virtual clients is also possible:

$$[S_1|D_1] = \begin{bmatrix} 0 & 44 & 12 & 0 \\ 1 & 44 & 8 & 14 \\ 3 & 57 & 0 & 4 \\ 4 & 46 & 0 & 0 \end{bmatrix}.$$

The first virtual client is obtained by aggregating clients 1 and 4, the second by aggregating clients 2 and 3, and the third by aggregating clients 2 and 5 from the original set of clients.

In the remainder of this section we will show that the second rule is equivalent to the Khumawala's "delta" rule (see [23, 43, 91]). In fact this rule is a special case of Cherenin's Excluding Rules, and our Preservation Rules (see Chap. 2), developed for supermodular functions (see [29]). In contrast to Khumawala [91] we use Hammer–Beresnev functions to justify the correctness of "delta" and "omega" rules, since such a justification leads to an efficient implementation of these rules.

The pseudo-Boolean representation of the SPLP allows us to develop rules using which we can peg certain variables in a solution by examining the coefficients of the Hammer–Beresnev functions. The rule that we use here is described in [68] as a Pegging Rule. We assume, without loss of generality, that the instance is not separable, i.e., we cannot partition I into sets I_1 and I_2, and J into sets J_1 and J_2, such that the transportation costs from sites in I_1 to clients in J_2, and from sites in I_2 to clients in J_1 are not finite. We also assume without loss of generality that the site indices are arranged in nonincreasing order of $f_i + \sum_{j \in J} c_{ij}$ values.

Theorem 4.2 (Cherenin's Excluding Rules, See Also Khumawala's "Delta" and "Omega" Rules [91] and Pegging Rule in [68]). *Let $\mathcal{H}_{[F|C]}(\mathbf{y})$ be the Hammer–Beresnev function corresponding to the SPLP instance $[F|C]$ in which like terms have been aggregated. Let $a_k = (\sum_{j:\pi_{1j}=k} \Delta c[1, j]) - f_k$ be the coefficient of the linear term corresponding to y_k and let*

$$t_k = \sum_{\substack{j=1 \\ j:k \in \{\pi_{1j},\dots,\pi_{pj}\}}}^{n} \sum_{p=2}^{m-1} \Delta c[p, j]$$

be the sum of the coefficients of all nonlinear terms containing y_k for each site index k. Then the following holds.

(a) RO: If $a_k \geq 0$, then there is an optimal solution $\mathbf{y}^\star$ in which $y_k^\star = 0$, else
(b) RC: If $a_k + t_k \leq 0$, then there is an optimal solution $\mathbf{y}^\star$ in which $y_k^\star = 1$, provided $y_i^\star \neq 1$ for some $i \neq k$.

Proof. (a) Suppose $a_k \geq 0$. Let us consider a solution $\mathbf{y}$ in which $y_k = 1$ and a solution $\mathbf{y}'$ in which $y_i' = y_i$ for each $i \neq k$, and $y_k' = 0$. Now $\mathcal{H}_{[F|C]}(\mathbf{y}) - \mathcal{H}_{[F|C]}(\mathbf{y}') \geq a_k \geq 0$. Hence $\mathbf{y}'$ is preferable to $\mathbf{y}$. This shows that $y_k = 0$ is an optimal solution.

(b) Next suppose that $a_k + t_k \leq 0$. Consider two solutions $\mathbf{y}$ and $\mathbf{y}'$, such that $y_i = y_i'$ for each $i \neq k$, $y_k = 0$, and $y_k' = 1$. Then

$$\mathcal{H}_{[F|C]}(\mathbf{y}') - \mathcal{H}_{[F|C]}(\mathbf{y})$$

$$= \left\{ \sum_{i=1}^{m} f_i(1 - y_i') + \sum_{j=1}^{n} \sum_{p=1}^{m-1} \Delta c[p, j] \prod_{r=1}^{p} y_{\pi_{rj}}' \right\}$$

$$- \left\{ \sum_{i=1}^{m} f_i(1 - y_i) + \sum_{j=1}^{n} \sum_{p=1}^{m-1} \Delta c[p, j] \prod_{r=1}^{p} y_{\pi_{rj}} \right\}$$

$$= \left\{ -f_k y_k' - \sum_{\substack{i=1 \\ i \neq k}}^{m} f_i y_i' + \underbrace{\sum_{j=1}^{n} \sum_{\substack{p=s+1 \\ s:\pi_{sj}=k}}^{m-1} \Delta c[p, j] \prod_{r=1}^{p} y_{\pi_{rj}}'}_{A} \right.$$

$$+ \underbrace{ \sum_{j=1}^{n} \sum_{p=1}^{r:\pi_{rj}=k} \sum_{p=1}^{p} \Delta c[p,j] \prod_{r=1}^{p} y'_{\pi_{rj}} \Bigg\} }_{B}$$

$$- \Bigg\{ \underbrace{ - f_k y_k - \sum_{\substack{i=1 \\ i \neq k}}^{m} f_i y_i }_{C} + \sum_{\substack{j=1 \\ j:k \in \{\pi_{1j},\ldots,\pi_{pj}\}}}^{n} \sum_{p=1}^{m-1} \Delta c[p,j] \prod_{r=1}^{p} y'_{\pi_{rj}} }$$

$$+ \underbrace{ \sum_{\substack{j=1 \\ j:k \notin \{\pi_{1j},\ldots,\pi_{pj}\}}}^{n} \sum_{p=1}^{m-1} \Delta c[p,j] \prod_{r=1}^{p} y_{\pi_{rj}} \Bigg\} }_{D} \qquad (4.14)$$

Notice that the terms marked A and C cancel each other since $y_i = y'_i$ when $i \neq k$, as do the terms marked B and D. Canceling these terms and setting $y_k = 0$ and $y'_k = 1$ in (4.14) we obtain

$$\mathcal{H}_{[F|C]}(\mathbf{y}') - \mathcal{H}_{[F|C]}(\mathbf{y})$$

$$= \Bigg\{ - f_k + \sum_{\substack{j=1 \\ j:k \in \{\pi_{1j},\ldots,\pi_{pj}\}}}^{n} \sum_{p=1}^{m-1} \Delta c[p,j] \prod_{r=1}^{p} y'_{\pi_{rj}} \Bigg\} \qquad (4.15)$$

which, on separating the linear and nonlinear terms

$$= \Bigg\{ a_k + \sum_{\substack{j=1 \\ j:k \in \{\pi_{1j},\ldots,\pi_{pj}\}}}^{n} \sum_{p=2}^{m-1} \Delta c[p,j] \prod_{r=1}^{p} y'_{\pi_{rj}} \Bigg\}. \qquad (4.16)$$

An upper bound to (4.16) is $a_k + t_k$, which is obtained by setting $y'_i = 1$ for each $i \in I$, since all nonlinear terms in the Hammer–Beresnev function have nonnegative coefficients. Thus

$$\mathcal{H}_{[F|C]}(\mathbf{y}') - \mathcal{H}_{[F|C]}(\mathbf{y}) \leq (a_k + t_k) \leq 0. \qquad (4.17)$$

Hence $\mathbf{y}'$ is preferable to $\mathbf{y}$. This shows that $y_k = 1$ in an optimal solution. Of course, if $y_i^* = 1$ for all $i \neq k$, then setting y_k^* to 1 would yield an infeasible solution. $\qquad \square$

Note that $t_k \geq 0$ for each index k, since the nonlinear terms of the Hammer–Beresnev function are nonnegative. Thus $a_k + t_k \leq 0$ implies that $a_k \leq 0$. If $t_k = 0$, then there is a possibility of a_k being equal to zero, but this possibility is taken care of in the first part of the rule.

The importance of the ordering of the site indices is demonstrated in the following lemma.

Lemma 4.3. *If $a_k < 0$ and $a_k + t_k \leq 0$ for each $k \in I$ in $\mathcal{H}_{[F|C]}(\mathbf{y})$ for the SPLP instance $[F|C]$, then an optimal solution would be $(1,1,\ldots,1,0)$ assuming that the site indices are arranged in nonincreasing order of $f_i + \sum_{j \in J} c_{ij}$ values.*

Proof. Let us initially relax the constraint $\mathbf{y} \neq \mathbf{1}$ in (4.11). In such a case it is easy to see that the optimal solution would be $\mathbf{y} = \mathbf{1}$ (from (4.17)). If we reimpose the constraint, we need to set one or more y_k values to 0. Changing $y_k = 0$ for any variable $k \in I$ increases the value of the Hammer–Beresnev function by $f_k + \sum_{j \in J} c_{kj}$. Note that setting $y_k = 0$ does not affect the non-positive nature of $a_i + t_i$, $i \neq k$, since this operation does not affect a_i and can only reduce the value of t_i. Also note that setting any additional variable y_i, $i \neq k$ to 0 cannot reduce the value of Hammer–Beresnev function since $a_i < 0$ and $a_i + t_i \leq 0$ for each $i \neq k$. The result follows. $\square$

The lemma above is illustrated by the following example. Consider an SPLP instance $[F|C]$, $m = n = 3$, in which $F = (99, 100, 98)$ and

$$C = \begin{bmatrix} 0 & 10 & 13 \\ 10 & 0 & 16 \\ 13 & 16 & 0 \end{bmatrix}$$

The Hammer–Beresnev function for this instance is

$$\mathcal{H}_{[F|C]} = 297 - 89y_1 - 90y_2 - 85y_3 + 9y_1y_2 + 3y_1y_3.$$

It is clear that $a_k < 0$ and $a_k + t_k < 0$ for $k = 1, 2$, and 3. Therefore the Pegging Rule will solve this instance completely, set $y_k = 1$ the first two sites it encounters, and set $y_k = 0$ for the last site. However, the solution would be correct only if the last site encountered has the lowest $f_i + \sum_{j \in J} c_{ij}$ value, i.e., if site 1 is considered after sites 2 and 3. In general therefore, the sites i should be ordered in nonincreasing values of $f_i + \sum_{j \in J} c_{ij}$.

Since there are $\mathcal{O}(mn)$ terms in the Hammer–Beresnev function corresponding to an SPLP instance with m candidate sites and n clients, the computational complexity of the preprocessing rule stated above is $\mathcal{O}(mn)$.

Notice that if at any preprocessing step, we can determine that $y_k = 1$ for a certain site k, then we need not include the row corresponding to site k in our calculations, and can therefore drop this row from the extended matrix in the succeeding steps. This deletion of rows is not possible if $y_k = 0$, since we do not know beforehand the whole set of clients that be served by a plant located at this site in any equivalent instance of the SPLP. The preprocessing rules also allow us to reduce the number of clients in the problem. If there is a client, the cost of satisfying whose demand by a site determined to be open by RO is less than the cost of satisfying it by any site whose status was not determined by preprocessing, then that client could be removed from further consideration. We could also cluster the clients based on the Hammer–Beresnev function, as illustrated in the following Example.

Example 4.3. Consider the SPLP instance

$$[F|C] = \begin{bmatrix} 7 & 7 & 15 & 10 & 7 & 10 \\ 3 & 10 & 17 & 8 & 11 & 22 \\ 3 & 16 & 7 & 6 & 18 & 14 \\ 6 & 11 & 7 & 6 & 12 & 8 \end{bmatrix}.$$

The Hammer–Beresnev function for this instance is $\mathcal{H}_{[F|C]}((y_1,y_2,\ y_3,y_4)) = 54 + 0y_1 - 3y_2 - 3y_3 - 4y_4 + 2y_1y_2 + 4y_1y_4 + 10y_3y_4 + 11y_1y_2y_4 + 10y_1y_3y_4 + 2y_2y_3y_4$. Since the coefficient of y_1 is zero we can set $y_1 = 0$. The Hammer–Beresnev function then becomes $\mathcal{H}_{[F|C]}((0,y_2,y_3,y_4)) = 54 - 3y_2 - 3y_3 - 4y_4 + 10y_3y_4 + 2y_2y_3y_4$. The coefficient of the linear term involving y_2 is negative and its magnitude in the revised Hammer–Beresnev polynomial is 3, while the sum of all terms containing y_2 in the transportation cost component is 2. So we can set $y_2 = 1$. The Hammer–Beresnev function then changes to $\mathcal{H}_{[F|C]}((0,1,y_3,y_4)) = 51 - 3y_3 - 4y_4 + 12y_3y_4$. One of the instances that such a Hammer–Beresnev function corresponds to is the following one (with rows corresponding to y_1, y_3 and y_4, respectively, since we have deleted the row corresponding to y_2).

$$[S|D] = \begin{bmatrix} 0 & 56 \\ 3 & 44 \\ 4 & 44 \end{bmatrix}.$$

It is easy to see that an optimal solution to this instance is $y_1 = y_3 = 0$ and $y_2 = y_4 = 1$. So an optimal solution to the SPLP instance is to set up plants at sites 1 and 3.

Hence we have reduced the size of the instance at hand, and in this case arrived at an optimal solution to the original instance using the preprocessing rules described above.

We carried out some preliminary computation to check the strength of our preprocessing rule. We used 12 benchmark problems in the OR-Library maintained by Beasley [10]. The results are summarized in column "m after Procedure a" of Table 4.1. Notice that the status of almost half of the number of sites could be predicted using the preprocessing rule. In particular, the second part of the rule, which allows us to predict sites which will *not* be opened in an optimal solution, is quite powerful for these instances.

4.4 Ingredients of Data Correcting for the SPLP

Data correcting is a method in which we alter the data in a problem instance to convert it to an instance that is easily solvable. This methodology was first introduced in [59]. In this subsection we illustrate the method for the SPLP when

Table 4.1 Number of free locations after preprocessing SPLP instances in the OR-Library

			m after procedure		
Problem	m	n	a	b	c
cap71	16	50	4	0	0
cap72	16	50	6	0	0
cap73	16	50	6	3	3
cap74	16	50	2	0	0
cap101	25	50	9	0	0
cap102	25	50	13	3	0
cap103	25	50	14	0	0
cap104	25	50	12	0	0
cap131	50	50	34	32	8
cap132	50	50	27	25	5
cap133	50	50	25	19	10
cap134	50	50	19	0	0

the instance data are represented by the fixed cost vector and the transportation cost matrix. However, it can be applied to a wide variety of optimization problems, and in particular, to the SPLP represented as a Hammer–Beresnev function.

Consider an instance $[F|C]$ of the SPLP. The objective of the problem is to compute a set P, $\emptyset \subset P \subseteq I$, that minimizes $f_{[F|C]}(P)$. Also consider an SPLP instance $[S|D]$ that is known to be polynomially solvable. Let $P^\star_{[F|C]}$ and $P^\star_{[S|D]}$ be optimal solutions to $[F|C]$ and $[S|D]$, respectively. Let us define the proximity measure $\rho([F|C],[S|D])$ between the two instances as

$$\rho([F|C],[S|D]) = \sum_{i \in I}|f_i - s_i| + \sum_{j \in J}\max\{|c_{ij} - d_{ij}| : i \in I\}. \qquad (4.18)$$

We use $\max\{|c_{ij} - d_{ij}| : i \in I\}$ in (4.18) instead of the expression $\sum_{i \in I}|c_{ij} - d_{ij}|$, since, in an optimal solution, the demand of each client is satisfied by a single facility, only one element in each column in the transportation matrix will contribute to the cost of the optimal solution.

Notice that $\rho([F|C],[S|D])$ is defined only when the instances $[F|C]$ and $[S|D]$ are of the same size. Also note that $\rho([F|C],[S|D])$ can be computed in time polynomial in the size of the two instances. The following theorem, which forms the basis for data correcting, shows that $\rho([F|C],[S|D])$ is an upper bound to the difference between the *unknown* optimal costs for the SPLP instances $[F|C]$ and $[S|D]$.

Theorem 4.3. *Let $[F|C]$ and $[S|D]$ be two SPLP instances of the same size, and let $P^\star_{[F|C]}$ and $P^\star_{[S|D]}$ be optimal solutions to $[F|C]$ and $[S|D]$, respectively. Then*

$$|f_{[F|C]}(P^\star_{[F|C]}) - f_{[S|D]}(P^\star_{[S|D]})| \leq \rho([F|C],[S|D]).$$

Proof. There are two cases to consider.

Case 1: $f_{[F|C]}(P^\star_{[F|C]}) \le f_{[S|D]}(P^\star_{P[S|D]})$, and
Case 2: $f_{[F|C]}(P^\star_{[F|C]}) > f_{[S|D]}(P^\star_{[S|D]})$. We only prove Case 1 here; the proof of Case 2 is similar to that of Case 1.

$$f_{[F|C]}(P^\star_{[F|C]}) - f_{[S|D]}(P^\star_{[S|D]}) \le f_{[F|C]}(P^\star_{[S|D]}) - f_{[S|D]}(P^\star_{[S|D]})$$

$$= \sum_{i \in P^\star_{[S|D]}} [f_i - s_i] + \sum_{j \in J} \left(\min\left\{ c_{ij} : i \in P^\star_{[S|D]} \right\} \right.$$

$$\left. - \min\left\{ d_{ij} : i \in P^\star_{[S|D]} \right\} \right).$$

Let $c_{i_c(j)j} = \min\{c_{ij} : i \in P^\star_{[S|D]}\}$ and $d_{i_d(j)j} = \min\{d_{ij} : i \in P^\star_{[S|D]}\}$. Then

$$f_{[F|C]}(P^\star_{[F|C]}) - f_{[S|D]}(P^\star_{[S|D]})$$

$$\le \sum_{i \in P^\star_{[S|D]}} [f_i - s_i] + \sum_{j \in J} [c_{i_c(j)j} - d_{i_d(j)j}]$$

$$\le \sum_{i \in P^\star_{[S|D]}} [f_i - s_i] + \sum_{j \in J} [c_{i_d(j)j} - d_{i_d(j)j}]$$

$$\le \sum_{i \in P^\star_{[S|D]}} [f_i - s_i] + \sum_{j \in J} [\max\{c_{ij} - d_{ij} : i \in P^\star_{[S|D]}\}]$$

$$\le \sum_{i \in P^\star_{[S|D]}} |f_i - s_i| + \sum_{j \in J} [\max\{|c_{ij} - d_{ij}| : i \in I\}]$$

$$\le \sum_{i \in I} |f_i - s_i| + \sum_{j \in J} [\max\{|c_{ij} - d_{ij}| : i \in I\}]$$

$$= \rho([F|C], [S|D]). \qquad \square$$

Theorem 4.3 implies that if we have an optimal solution to an SPLP instance $[S|D]$, then we have an upper bound for *all* SPLP instances $[F|C]$ of the same size. This upper bound is actually the distance between the two instances, distances being defined by the proximity measure (4.18). Also if the solution to $[S|D]$ can be computed in polynomial time (i.e., $[S|D]$ belongs to a polynomially solvable special case), then an upper bound to the cost of an *as yet unknown* optimal solution to $[F|C]$ can be obtained in polynomial time. If the distance between the instances is not more than a prescribed accuracy ε, then the optimal solution of $[S|D]$ is, in fact, a solution to $[F|C]$ within the prescribed accuracy. This theorem forms the basis of data correcting.

In general, the data correcting procedure works as follows. It assumes that we know a class of polynomially solvable instances of the problem. It starts by choosing a polynomially solvable SPLP instance $[S|D]$ from that class of instances, preferably as close as possible to the original instance $[F|C]$. If $\rho([F|C], [S|D]) \leq \varepsilon$, the procedure terminates and returns an optimal solution to $[S|D]$ as an approximation of an optimal solution to $[F|C]$. The instance $[F|C]$ is said to be "corrected" to the instance $[S|D]$, which is solved polynomially to generate the solution output by the procedure. Otherwise, the set of feasible solutions for the problem is partitioned into two subsets. For the SPLP, one of these subsets is comprised of solutions that locate a plant at a given site, and the other is comprised of solutions that do not. The two new instances thus formed are perturbed in a way that they both change into instances that are within a distance ε from a polynomially solvable instance. The procedure is continued until an instance with a proximity measure not more than ε is obtained for all the subsets generated.

4.4.1 The Reduction Procedure

The Data Correcting Algorithm (DCA) that we propose in this chapter is the one that uses a strong Reduction Procedure (RP) to reduce the original instance into a smaller "core" instance, and then uses a data correcting procedure (DCP, see Fig. 3.1 in Sect. 3.2) to obtain a solution to the original instance, whose cost is not more than a prespecified amount ε more than the cost of an optimal solution.

The first preprocessing RO and RC rules (see Theorem 4.2 in Sect. 4.3) for the SPLP involving both fixed costs and transportation costs appeared in [3, 29, 46, 89, 91].

Notice that RO and RC primarily try to either open or close sites. If it succeeds, it also changes the Hammer–Beresnev function for the instance, reducing the number of nonlinear terms therein. In the remaining portion of this subsection, we describe a completely new *Reduction Procedure* (RP), whose primary aim is to reduce the coefficients of terms in the Hammer–Beresnev function, and if we can reduce it to zero, to eliminate the term from the Hammer–Beresnev function. This procedure is based on fathoming rules of BnB algorithms and data correcting principles.

Let us assume that we have an upper bound (UB) on the cost of an optimal solution for the given SPLP instance. This can be obtained by running a heuristic on the problem data. Now consider any nonlinear term $s \prod_{r=1}^{k} y_{\pi_{rj}}$, in the Hammer–Beresnev function. This term will contribute to the cost of a solution, only if plants are *not* located in any of the sites $\pi_{1j}, \ldots, \pi_{kj}$. Let lb be a lower bound on the optimal solution of the SPLP with respect to the subspace for which no facilities are located in sites $\pi_{1j}, \ldots, \pi_{kj}$. If lb $\leq$ UB, then we cannot make any judgement about this term. On the other hand, if lb $>$ UB, then we know that there cannot be an optimal solution with $y_{\pi_{1j}} = \cdots = y_{\pi_{kj}} = 1$. In this case, if we reduce the coefficient s by lb $-$ UB $- \varepsilon$, ($\varepsilon > 0$, small), then the new Hammer–Beresnev function and the original one have identical sets of optimal solutions. Note that the values of upper UB and lower lb bounds are calculated not necessarily for the same subspace of feasible solutions. If after the reduction, s is non-positive, then the term can be removed from the Hammer–Beresnev function. Such changes in the Hammer–Beresnev function alter the values of t_k, and can possibly allow us to use Cherenin's Excluding Rules to close certain sites. Once some sites are closed, some of the linear terms in the Hammer–Beresnev function change into constant terms, and some of the quadratic terms change into linear ones. These changes cause changes in both the a_k and the t_k values, and can make further application of Cherenin's Excluding Rules, thus preprocessing some other sites, and making further changes in the Hammer–Beresnev function. A pseudocode of the reduction procedure $\mathrm{RP}(\mathscr{H}_{[F|C]}(\mathbf{y}))$ is provided below.

Procedure $\mathrm{RP}(\mathscr{H}_{[F|C]}(\mathbf{y}))$
begin
    repeat
        Compute an upper bound UB for the instance;
        for each nonlinear term $s\prod_{r=1}^{k} y_{\pi_{rj}}$ in $\mathscr{H}_{[F|C]}(\mathbf{y})$ **do**
        begin
            Compute lower bound lb on the cost of solutions in
            which plants are not located in sites $\pi_{1j},\dots,\pi_{kj}$;
            if lb > UB **then**
                Reduce the coefficient of the term by
                $\max\{s, \mathrm{lb} - \mathrm{UB} - \varepsilon\}$;
        end
        Apply Khumawala's rules until no further preprocessing
        is possible;
        Recompute the Hammer–Beresnev function $\mathscr{H}_{[F|C]}(\mathbf{y})$;
    until no further preprocessing of sites was achieved
    in the current iteration;
end;

Let us consider the application of all preprocessing rules to the following SPLP instance:

$$[F|C] = \begin{bmatrix} 9 & 7 & 12 & 22 & 13 \\ 4 & 8 & 9 & 18 & 17 \\ 3 & 16 & 17 & 10 & 27 \\ 6 & 9 & 13 & 10 & 11 \end{bmatrix}. \tag{4.19}$$

Two possible ordering matrices corresponding to C are

$$\Pi_1 = \begin{bmatrix} 1 & 2 & 3 & 4 \\ 2 & 1 & 4 & 1 \\ 4 & 4 & 2 & 2 \\ 3 & 3 & 1 & 3 \end{bmatrix} \text{ and } \Pi_2 = \begin{bmatrix} 1 & 2 & 4 & 4 \\ 2 & 1 & 3 & 1 \\ 4 & 4 & 2 & 2 \\ 3 & 3 & 1 & 3 \end{bmatrix}. \tag{4.20}$$

The Hammer–Beresnev function is $\mathscr{H}_{[F|C]}(\mathbf{y}) = [9(1-y_1)+4(1-y_2)+3(1-y_3)+6(1-y_4)] + [7+1y_1+1y_1y_2+7y_1y_2y_4] + [9+3y_2+1y_1y_2+4y_1y_2y_4] + [10+$

$0y_3 + 8y_3y_4 + 4y_2y_3y_4] + [11 + 2y_4 + 4y_1y_4 + 10y_1y_2y_4] = 59 - 8y_1 - y_2 - 3y_3 - 4y_4 + 2y_1y_2 + 4y_1y_4 + 8y_3y_4 + 21y_1y_2y_4 + 4y_2y_3y_4$.

The values of a_k, t_k and $a_k + t_k$ are as follows:

k	1	2	3	4
a_k	−8	−1	−3	−4
t_k	27	27	12	37
$a_k + t_k$	19	26	9	33

It is clear that neither RO nor RC is applicable here, since the coefficient of the term $21y_1y_2y_4$ is too large. Therefore, we try to reduce this coefficient by applying RP.

The upper bound UB $= 51$ to the original problem can be obtained by setting $y_1 = y_4 = 1$ and $y_2 = y_3 = 0$. A lower bound to the subproblem under the restriction $y_1 = y_2 = y_4 = 1$ is 73, since $\mathcal{H}_{F|C}(1,1,0,1) = 73$. In virtue of RP, we can reduce the coefficient of $21y_1y_2y_4$ by $73 - 51 - \varepsilon = 20$, so that the new Hammer–Beresnev function with the same set of optimal solutions as the original function becomes, $\mathcal{H}'(y) = 59 - 8y_1 - y_2 - 3y_3 - 4y_4 + 2y_1y_2 + 4y_1y_4 + 8y_3y_4 + 1y_1y_2y_4 + 4y_2y_3y_4$. The updated values of a_k, t_k, and $a_k + t_k$ are presented below. RC can immediately

k	1	2	3	4
a_k	−8	−1	−3	−4
t_k	7	7	12	17
$a_k + t_k$	−1	6	9	13

be applied in this situation to set $y_1 = 1$. Updating $\mathcal{H}'(y)$, we can apply RO and set $y_2 = y_4 = 0$. This allows us to apply RC again to set $y_3 = 1$, yielding the optimal solution $(1,0,1,0)$ with cost 48.

4.4.2 The Data Correcting Procedure

We have used our data correcting procedure **Procedure** DC$(S,T,\varepsilon;\lambda,\gamma)$ from Chap. 3 in which we set $\delta^+ = \max\{a_k \mid k \in S \setminus T\}$ and $\delta^- = \max\{a_k + t_k \mid k \in S \setminus T\}$. Let us suppose that the DC procedure is applied to the SPLP instance $[F|C]$. On termination, it outputs two subsets S and T, $\emptyset \subset S \subseteq T \subseteq I$. If $S = T$, then the instance is said to have been solved by this procedure, and set S is an optimal solution. Since the PP procedure is a polynomial time algorithm, instances that it solves to optimality constitute a class of algorithmically defined polynomially solvable instances. We have called such instances *PP-solvable*. We use this class of polynomially solvable instances in our algorithm, since it is one of the best among the polynomially solvable cases discussed in [64].

Next suppose that the given instance is not PP-solvable. In that case we try to extend the idea of the PP procedure to obtain a solution such that the difference between its cost and the cost of an optimal solution is bounded by a predefined value ε. This is the basic idea behind the data correcting procedure.

In case $\delta = \min(\delta^-, \delta^+) > \varepsilon$, then data correction cannot guarantee a solution within the prescribed allowable accuracy, and hence we need to use a branching procedure.

The data correcting procedure (DCP, see below) in our algorithm takes two sets $S, T \subseteq I$ ($\emptyset \subset S \subset T \subseteq I$) and ε as input. It outputs a solution λ and a bound γ, such that $f_{[F|C]}(\lambda) - f_{[F|C]}(P^\star) \leq \gamma \leq \varepsilon$, where $P^\star$ is an optimal solution to $[F|C]$. It is a recursive procedure that first tries to reduce the set $T \setminus S$ by applying Lemma 3.1b and a. If Lemma 3.1b and a cannot be applied, then it tries to apply Lemma 3.1d and c to reduce $T \setminus S$ (see Sect. 3.2). We do not use the reduction procedure at this stage since it increases the computational times substantially without reducing the core problem appreciably. If even these lemmas cannot be applied, then the procedure branches on a member $k \in T \setminus S$ and invokes two instances of DCP, one with sets $S \cup \{k\}$ and T, and the other with sets S and $T \setminus \{k\}$. Notice that the solutions searched by the two invocations of DCP are mutually exclusive and exhaustive. A bound is used to remove unpromising subproblems from the solution tree. The choice of the branching variable $k \in T \setminus S$ in DCP is motivated by the observation that $a_k < 0$ and $t_k + a_k > 0$ for each of these indices. (These are the preconditions of the branching rule.) A plant would have been located in this site in an optimal solution if the coefficient of the linear term involving y_k in the Hammer–Beresnev function had been increased by $-a_k$. We could have predicted that a plant would not be located there if the same coefficient had been decreased by $t_k + a_k$. Therefore we could use $\phi_k = \text{average}(-a_k, t_k + a_k) = \frac{t_k}{2}$ as a measure of the chance that we will *not* be able to predict the fate of site k in any subproblem of the current subproblem. If we want to reduce the size of the BnB tree by assigning values to such variables, then we can think of a branching function (see [67]) that branches on the index $k \in P_U \setminus P_L$ with the largest ϕ_i value.

4.5 Computational Experiments

The execution of the DCA can be divided into two stages: a *preprocessing* stage in which the given instance is reduced to a core instance by using RP; and a *solution* stage in which the core instance is solved using DCP.

In the preprocessing stage we experimented with the following three reduction procedures:

(a) The "delta" and "omega" rules from [91].
(b) Procedure RP with the combinatorial Khachaturov–Minoux bound to obtain a lower bound.
(c) Procedure RP with the LP dual-ascent Erlenkotter bound (see [44]) to obtain a lower bound.

Table 4.2 Number of nonzero nonlinear terms in the Hammer–Beresnev function after preprocessing SPLP instances in the OR-Library

	# of nonzero terms			
		After procedure		
Problem	Before preprocessing	a	b	c
cap71	699	6	0	0
cap72	699	12	0	0
cap73	699	13	2	2
cap74	699	1	0	0
cap101	1,147	24	0	0
cap102	1,147	33	2	0
cap103	1,147	38	0	0
cap104	1,147	29	0	0
cap131	2,389	163	135	8
cap132	2,389	112	92	3
cap133	2,389	101	60	11
cap134	2,389	62	0	0

The Khachaturov–Minoux bound lb is a combinatorial bound for general supermodular functions (see Lemma 3.2 due to [89, 105]).

We also experimented with the Khachaturov–Minoux bound and the Erlenkotter bound in the implementation of the DCP.

The effectiveness of the reduction procedure can be measured either by computing the number of free locations in the core instance, or by computing the number of nonzero nonlinear terms present in the Hammer–Beresnev function of the core instance. Note that the number of nonzero nonlinear terms present in the Hammer–Beresnev function is an upper bound on the number of unassigned customers in the core instance. Tables 4.1 and 4.2 show how the various methods of reduction perform on the benchmark SPLP instances in the OR-Library [10]. The existing preprocessing rules due to [29, 91] (i.e., procedure (a), which was used in the SPLP example in [66]) cannot solve any of the OR-Library instances to optimality. However, the variants of the new RP [i.e., procedures (b) and (c)] solve a large number of these instances to optimality. Procedure (c) based on the Erlenkotter bound is marginally better than procedure (b) in terms of the number of free locations (Table 4.1), but substantially better in terms of the number of nonzero nonlinear terms in the Hammer–Beresnev function (Table 4.2).

The information in Tables 4.1 and 4.2 can be combined to show that some of the problems that are not solved by these procedures can actually be solved by inspection of the core instances. For example, consider cap74. We see that the core problem [using procedure (a)] has two free variables and one nonlinear term. Therefore the Hammer–Beresnev function of the core instance looks like

$$A + py_u + qy_w + ry_uy_w,$$

where $p, q < 0, r > 0, \min\{p+r, p+q\} > 0$ and A is a constant. The minima of such functions are easy to obtain by inspection.

Table 4.3 Comparison of bounds used with the DCA on Körkel-type instances with $m = n = 65$

Problem set	Execution time of the DCP (s)	
	Khachaturov–Minoux bound	Erlenkotter bound
Set 1	119.078	0.022
Set 2	290.388	0.040
Set 3	458.370	0.056
Set 4	158.386	0.054
Set 9	428.598	0.588
Set 10	542.530	0.998
Set 11	479.092	2.280

In addition, Tables 4.1 and 4.2 demonstrate the superiority of the new preprocessing rule over the "delta" and "omega" rules. Consider for example the problem cap132. The "delta" and "omega" rules reduce the problem size from $m = 50$ and 2,389 nonzero nonlinear terms to $m' = 27$ and 112 nonzero nonlinear terms. However, the new preprocessing rule reduces the same problem to one having $m' = 5$ and 3 nonzero nonlinear terms.

In order to test the effect of bounds in the DCA, we compared the execution times of DCA using the two bounds on some difficult problems of the type suggested in [93] (see Sect. 4.5.4 for more details). The problems were divided into seven sets. Each set consists of five problems, each having 65 sites and 65 clients (see Sect. 4.5.4 for more details regarding these problems). From Table 4.3 we see that the Erlenkotter bound reduces the execution time taken by the Khachaturov–Minoux bound (that was used in the SPLP example in [66]) by a factor of more than 100. This is not surprising, since the Khachaturov–Minoux bound is derived for a general supermodular function, while the Erlenkotter bound is specific to the SPLP.

We report our computational experience with the DCA on several benchmark instances of the SPLP in the remainder of this section. The performance of the algorithm is compared with that of the algorithms described in the chapters that suggested these instances. We implemented the DCA in PASCAL, compiled it using Prospero Pascal, and ran it on a 733 MHz Pentium III machine. The computation times we report are in seconds on our machine.

4.5.1 Bilde and Krarup-Type Instances

These are the earliest benchmark problems that we consider here. The exact instance data are not available, but the process of generating the problem instances is described in [15]. There are 22 different classes of instances, and Table 4.4 summarizes their characteristics.

Table 4.4 Description of the instances in [15]

Type	m	n	f_i	c_{ij}
B	50	100	Uniform(1,000, 10,000)	Uniform(0, 1,000)
C	50	100	Uniform(1,000, 2,000)	Uniform(0, 1,000)
D^a_q	30	80	Identical, $1,000 \times q$	Uniform(0, 1,000)
E^a_q	50	100	Identical, $1,000 \times q$	Uniform(0, 1,000)

$^a q = 1, \ldots, 10$

Table 4.5 Results from Bilde and Krarup-type instances

Problem type	DCA		Bilde and Krarup [15]	
	Branching	CPU time	Branching	CPU time[a]
B	11.72	0.67	43.3	4.33
C	17.17	14.81	⋆	>250
D1	13.80	0.65	216	11
D2	12.13	0.38	218	24
D3	10.87	0.19	169	19
D4	10.25	0.15	141	17
D5	9.24	0.07	106	14
D6	8.99	0.09	101	15
D7	8.79	0.09	83	13
D8	8.60	0.09	55	11
D9	8.15	0.07	47	11
D10	7.29	0.03	43	11
E1	18.66	35.28	1,271	202
E2	16.14	8.64	1,112	172
E3	14.59	3.81	384	82
E4	13.65	2.74	258	65
E5	12.73	2.01	193	53
E6	11.82	0.90	136	43
E7	10.82	0.53	131	42
E8	10.79	0.68	143	48
E9	10.62	0.76	117	44
E10	10.36	0.69	79	37

[a] IBM 7,094 s

⋆ Could not be solved in 250 s

In our experiments we generated ten instances for each of the types of problems, and used the mean values of our solutions to evaluate the performance of our algorithm with the one used in [15]. In our implementation, we used the reduction procedure (b) and the Khachaturov–Minoux bound in the DCP.

The reduction procedure was not useful for these instances, but the DCA could solve all the instances in reasonable time. The results of our experiments are presented in Table 4.5. The performance of the algorithm implemented in [15] was measured in terms of the number of branching operations performed by the algorithm and its execution time in CPU seconds on a IBM 7094 machine.

Table 4.6 Description of the instances in [47]

Problem size	Density	Fixed costs' parameters	
$(m = n)$	δ	Mean	Standard deviation
10	0.300	4.3	2.3
20	0.150	9.4	4.8
30	0.100	13.9	7.4
50	0.061	25.1	14.1
70	0.043	42.3	20.7
100	0.025	51.7	28.9
150	0.018	186.1	101.5
200	0.015	149.5	94.4

We estimate the number of branching operations by our algorithm as the logarithm (to the base 2) of the number of subproblems it generated. From the table we see that the DCA reduces the number of subproblems generated by the algorithm in [15] by several orders of magnitude. This is especially interesting because Bilde and Krarup use a bound (discovered in 1967) identical to the Erlenkotter bound in their algorithm (see [93]) and we use the Khachaturov–Minoux bound. The CPU time required by the DCA to solve these problems was too low to warrant the use of any $\varepsilon > 0$.

4.5.2 Galvão and Raggi-Type Instances

Galvão and Raggi [47] developed a general 0–1 formulation of the SPLP and presented a three-stage method to solve it. The benchmark instances suggested in this work are unique, in that the fixed costs are assumed to come from a Normal distribution rather than the more commonly used Uniform distribution. The transportation costs for an instance of size $m \times n$ with $m = n$ are computed as follows. A network with a given arc density δ is first constructed, and the arcs in the network are assigned lengths sampled from a uniform distribution in the range $[1, n]$ (except for $n = 150$, where the range is $[1, 500]$). The transportation cost from i to j is the length of the cheapest path from i to j. The problem characteristics provided in [47] are summarized in Table 4.6.

As with the data in [15], the exact data for the instances are not known. So we generated ten instances for each problem size and used the mean values of the solutions for comparison purposes. In our DCA implementation, we used reduction procedure (b) and the Khachaturov–Minoux bound in the DCP. The comparative results are given in Table 4.7. Since the computers used are different, we cannot make any comments on the relative performance of the solution procedures. However, since the average number of subproblems generated by the DCA is always less than 10 for each of these instances, we can conclude that these problems are easy for our algorithm. In fact they are too easy for the DCA to warrant $\varepsilon > 0$.

Table 4.7 Results from Galvão and Raggi-type instances

Problem size $(m = n)$	DCA				Galvão and Raggi [47]	
	# solved by pre- processing	# of sub- problems[a]	CPU time[a]	# of open plants[a]	CPU time[b]	# of open plants
10	6	2.3	<0.001	4.7	<1	3
20	5	2.4	<0.001	9.0	<1	8
30	7	1.8	0.002	13.6	1	11
50	7	2.6	0.002	20.3	2	20
70	2	3.8	0.004	28.8	6	31
100	3	3.5	0.011	41.1	6	44
150	1	7.8	0.010	64.4	25	74
200	4	2.9	0.158	81.8	63	84

[a]Average over ten instances
[b]IBM 4,331 s

Note that the average number of opened plants in the optimal solutions to the instances we generated is quite close to the number of opened plants in the optimal solutions reported in [47]. Also observe that the reduction procedure was quite effective—it solved 35 of the 80 instances generated.

4.5.3   Instances from the OR-Library

The OR-Library [10] has a set of instances of the SPLP. These instances were solved in [9] using an algorithm based on the Lagrangian heuristic for the SPLP. Here too, we used reduction procedure (b) and the Khachaturov–Minoux bound in the DCP. We solved the problems to optimality using the DCA. The results of the computations are provided in Table 4.8. The execution times suggest that the DCA is faster than the Lagrangian heuristic described in [9]. The reduction procedure was also quite effective for these instances, solving 4 of the 16 instances to optimality, and reducing the number of free sites appreciably in the other instances. Once again the use of $\varepsilon > 0$ cannot be justified, considering the execution times of the DCA.

4.5.4   Körkel-Type Instances with 65 Sites

Korkel [93] described several relatively large Euclidean SPLP instances ($m = n = 100$, and $m = n = 400$) and used a BnB algorithm to solve these problems. The bound used in that work is an improvement on a bound based on the dual of the linear programming relaxation of the SPLP due to [44] and is extremely effective. The bound due to [44] is very effective because, for a large majority of SPLP instances, the optimal solution to the dual of the linear programming relaxation

Table 4.8 Results from OR-Library instances

Problem name	m	n	DCA m after pre-processing	# of sub-problems	CPU time	CPU time [9][a]	# of open plants
cap71	16	50	⋆	0	<0.01	0.11	11
cap72	16	50	⋆	0	<0.01	0.08	9
cap73	16	50	⋆	0	<0.01	0.11	5
cap74	16	50	⋆	0	<0.01	0.05	4
cap101	25	50	9	6	<0.01	0.18	15
cap102	25	50	13	16	<0.01	0.16	11
cap103	25	50	14	16	<0.01	0.14	8
cap104	25	50	12	7	0.01	0.11	4
cap131	50	50	34	196	0.01	0.31	15
cap132	50	50	27	183	0.02	0.28	11
cap133	50	50	25	71	<0.01	0.29	8
cap134	50	50	19	25	<0.01	0.15	4

[a]Cray-X-MP/28 s
⋆Instance solved by preprocessing only

Table 4.9 Description of the fixed costs for instances in [93]

Problem set	# of instances	Fixed cost for ith instance
Set 1	5	Identical, set at $141 + 6.6i$
Set 2	5	Identical, set at $174 + 6.6i$
Set 3	5	Identical, set at $207 + 6.6i$
Set 4	5	Identical, set at $174 + 66i$
Set10	5	Identical, set at $7,170 + 660i$
Set11	5	Identical, set at $7,120.5 + 333.3i$
Set12	5	Identical, set at $8,787 + 333.3i$

of the SPLP is integral. In this subsection, we use instances that have the same cost structure as the ones in [93] but for which $m = n = 65$. Instances of this size were not dealt with in [93]. We used reduction procedure (b) for the RP, and the Khachaturov–Minoux bound in the DCP.

In [93], 120 instances of each problem size are described. These can be divided into 28 sets (the first 18 sets contain 5 instances each, and the rest contain 3 instances each). We solved all the 120 instances we generated and found out that the instances in sets 1, 2, 3, 4, 10, 11, and 12 are more difficult to solve than others. We therefore used these instances in the experiments in this section. The transportation cost matrix for a Körkel instance of size $n \times n$ is generated by distributing n points in random within a rectangular area of size $700 \times 1,300$ and calculating the Euclidean distances between them. The fixed costs are computed as in Table 4.9.

The values of the results that we present for each set is the average of the values obtained for all the instances in that set. Interestingly, the preprocessing rules were found to be totally ineffective for all of these problems. Since the fixed costs are identical for all the sites, the sites are distributed randomly over a region, and

Table 4.10 Costs of solutions output by the DCA on Körkel-type instances with 65 sites

Problem set	Optimal	Acceptable accuracy[a]				
		1 %	2 %	3 %	5 %	10 %
Set 1	6,370.0	6,404.8	6,450.6	6,480.6	6,569.2	6,781.0
Set 2	6,920.6	6,952.2	6,971.4	7,028.4	7,123.8	7,320.2
Set 3	7,707.4	7,738.0	7,770.2	7,797.6	7,854.6	8,053.8
Set 4	9,601.2	9,642.4	9,680.2	9,698.4	9,786.6	9,932.0
Set10	146,691.2	146,896.6	146,909.6	147,543.6	148,062.0	151,542.2
Set11	168,598.4	168,858.2	169,655.0	170,341.6	170,597.0	173,913.8
Set12	186,386.3	186,729.7	187,112.0	188,002.7	188,854.2	192,528.7

[a] As a percentage of the optimal cost

Table 4.11 Execution times for the DCA on Körkel-type instances with 65 sites

Problem set	Optimal	Acceptable accuracy[a]				
		1 %	2 %	3 %	5 %	10 %
Set 1	119.078	90.948	70.758	55.494	43.200	20.426
Set 2	290.388	225.108	172.422	145.828	96.240	36.966
Set 3	458.370	339.420	259.022	203.036	150.216	50.378
Set 4	158.386	129.694	109.754	89.666	65.548	30.058
Set 10	428.598	370.120	319.804	283.832	230.078	142.090
Set 11	542.530	476.350	418.628	408.594	290.338	160.744
Set 12	479.092	416.472	370.832	326.572	261.835	149.038

[a] As a percentage of the optimal cost

the variable cost matrix is symmetric, no site presents a distinct advantage over any other. This prevents our reduction procedure to open or close any site. Table 4.10 shows the variation in the costs of the solution output by the DCA with changes in ε, and Table 4.11 shows the corresponding decrease in execution times.

The effect of varying the acceptable accuracy ε on the cost of the solutions output by the DCA is also presented graphically in Fig. 4.1. We define the *achieved accuracy* β as

$$\beta = \frac{\text{cost of solution output by the DCA} - \text{cost of optimal solution}}{\text{cost of optimal solution}}$$

and the *relative time* τ as

$$\tau = \frac{\text{execution time for the DCA for acceptable accuracy } \varepsilon}{\text{execution time for the DCA to compute an optimal solution}}$$

Note that the achieved accuracy β varies almost linearly with ε, with a slope close to 0.5. Also note that the relative time τ of the DCA reduces with increasing ε. The reduction is slightly better than linear, with an average slope of -8.

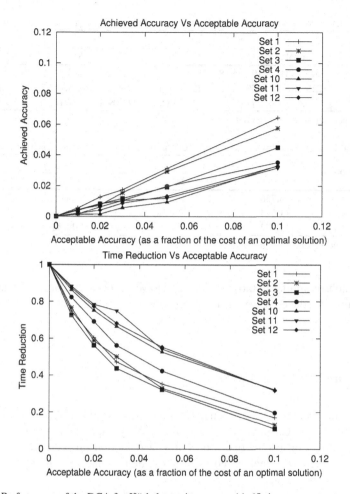

Fig. 4.1 Performance of the DCA for Körkel-type instances with 65 sites

4.5.5 Körkel-Type Instances with 100 Sites

We solved the benchmark instances in [93] with $m = n = 100$ to optimality and observed that the instances in sets 10–12 required relatively longer execution times. So we restricted further computations to instances in those sets. The fixed and transportation costs for these problems are computed in the procedure described in Sect. 4.5.4. Tables 4.12 and 4.13 show the results obtained by running the DCA on these problem instances. In our DCA implementation for solving these instances, we used reduction procedure (c) and the Erlenkotter bound in the DCP.

Figure 4.2 illustrates the effect of varying the acceptable accuracy ε on the cost of the solutions output by the DCA for the instances mentioned above. The nature

Table 4.12 Costs of solutions output by the DCA on Körkel-type instances with 100 sites

Problem set	Optimal	Acceptable accuracy[a]				
		1 %	2 %	3 %	5 %	10 %
Set 10	190,782.0	191,550.8	192,755.4	192,080.6	195,983.2	203,934.2
Set 11	219,583.4	220,438.8	222,393.6	221,947.2	228,467.2	235,963.4
Set 12	240,402.4	241,609.6	243,336.8	244,209.4	247,417.6	259,168.6

[a] As a percentage of the optimal cost

Table 4.13 Execution times for the DCA on Körkel-type instances with 100 sites

Problem set	Optimal	Acceptable accuracy[a]				
		1 %	2 %	3 %	5 %	10 %
Set 10	133.746	91.774	65.99	65.908	44.2	32.074
Set 11	81.564	55.356	39.554	38.348	33.628	17.598
Set 12	111.272	85.858	65.608	55.928	61.758	33.014

[a] As a percentage of the optimal cost

of the graphs is similar to those in Fig. 4.1. However, in several of the instances we noticed that β reduced when ε is increased, and in some other instances τ increased when ε was increased.

4.6   Concluding Remarks

In this chapter we tailor the general data correcting algorithm (DCA) for supermodular functions (see Chap. 3 and [66]) to the simple plant location problem (SPLP). This algorithm consists of two procedures, a reduction procedure to reduce the original instance to a smaller "core" instance, and a data correcting procedure to solve the core instance.

Theorem 4.3 can be considered as the basis of data correcting. It states that for two different instances of the SPLP of the same size, the difference between the costs of the *unknown* optimal solutions for these instances is bounded by a polynomially calculated distance between these instances. This distance is used to *correct* one of these instances in an implicit way by just *correcting* the value of the given accuracy parameter in the DCA.

An important contribution of this chapter is a new reduction procedure (RP), which when implemented in the DCA yields to a substantial reduction in the size of the original instance. This reduction procedure is much more powerful than the "delta" and "omega" reduction rules in [91]. It also incorporates the Erlenkotter bound specific to the SPLP (see [44]), which is more computationally efficient than the Khachaturov–Minoux bound used in [66]. The strength of the new RP based on the Erlenkotter bound is made obvious by the observation that none of the instances in the OR-Library could be solved by the "delta" and "omega" rules to optimality, but the new reduction procedure solves 75 % of them to optimality, and preprocesses

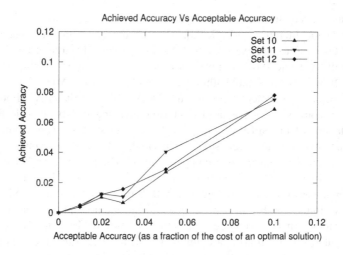

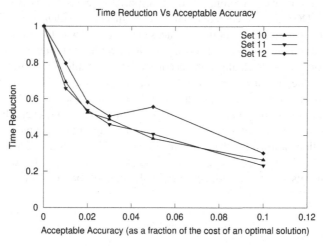

Fig. 4.2 Performance of the DCA for Körkel-type instances with 100 sites

at least twice the number of sites as the "delta" and "omega" rules for the remaining 25 % of the instances. Another contribution of the chapter is the incorporation of the Erlenkotter bound to the recursive BnB type data correcting procedure.

We have compared the performance of the Erlenkotter bound implemented in an usual BnB type algorithm (see [15]) and the Khachaturov–Minoux bound implemented in the DCP for the new RP and for fathoming subproblems created by the DCP. On the instances in [15], the number of subproblems created by the BnB type algorithm with Erlenkotter bound is found to be more than 1,000 times the number of subproblems created by the DCP based on the Khachaturov–Minoux bound.

We have tested the DCA on a broad range of different classes of instances available in the literature [15, 47, 93]. The striking computational result is the ability of the DCA to find exact solutions for many relatively large instances within fractions of a second. For example, an exact global optimum of the 200×200 instances from [47] was found within 0.2 s on a PC with a 733 MHz processor.

In all of our implementations for the DCA with Khachaturov–Minoux and Erlenkotter bounds, we have used data structures induced by pseudo-Boolean representations of the SPLP due to [80]. These data structures are conducive to efficient updating for the current subproblems in the DCA and sometimes show that a current subproblem remaining after application of the new RP has relatively small numbers of linear and nonlinear terms in the corresponding Hammer–Beresnev function and therefore can be solved by any BnB type algorithm for the SPLP.

We have found that for all instances in [93] the "delta" and "omega" reduction rules were totally ineffective since none of the sites presented any distinct advantage over any other (the fixed costs are almost identical for all sites, the sites are distributed randomly over a region, and the transportation costs matrix is symmetric). Anyway, the DCA has solved to optimality all the instances with $m = n = 100$ within fractions of a second except for the instances in sets 10–12 which required relatively longer execution times. On these sets of instances we have studied the behavior of the execution time and calculated the accuracy for acceptable values of ε. When the acceptable value of ε increases, we see that the costs of the solutions output by the DCA generally worsen, but the execution times also decrease.

In summary, our computational experience with the DCA on several benchmark instances known in the literature suggests that the algorithm compares well with other algorithms known for the problem. However, like any other BnB algorithm, DCA depends heavily on the quality of the bounds used. We believe that this algorithm merits serious consideration as a solution tool for the SPLP.

Chapter 5
Summary

In this book we study a class of algorithms for solving NP-hard problems called *data correcting* algorithms. A data correcting (DC) algorithm is a branch-and-bound type algorithm, in which the data of a given problem is "heuristically corrected" at the various stages in such a way that the new instance will be polynomially solvable and its optimal solution is within a prespecified deviation (called *prescribed accuracy*) from the optimal solution to the original problem.

The DC approach is applied to determining exact and approximate global optima of NP-hard problems. A DC algorithm consists of the following ingredients:

(a) A *polynomially solvable special case* (PSC) related to the original problem.
(b) A *proximity measure* between two problem instances, being a polynomially computable measure for the distance between the two instances. This measure provides an upper bound for the difference between the objective function values of the optimal solutions to the two instances.

The DC approach is based on the well-known literature in the polynomially solvable special cases. In the DC approach we can directly start with computational experiments based on the well-known literature in the polynomially solvable special cases. The choice of a branching element in the data-correcting approach is based on an element $e_k \in s_F \setminus s_R$ with the maximum contribution into the current value of a proximity measure, and hence leads to the reducing of the current value of proximity measure. Moreover, the values of proximity measure computed for different subproblems made useful any additional heuristic included in a DC algorithm. For example, the patching heuristic finds a feasible solution to the ATSP and provides an upper bound for the optimal value in any DC algorithm. In the DC algorithm the patching operation is used not only for finding a feasible solution to the ATSP but helps us to form a corrected instance (which has the patched solution as an optimal solution). We then use this corrected instance not only to compute an upper bound (see Theorem 1.2) of the cost difference between the patched solution and the *yet unknown* optimal solution to the original problem, but also to decide by which arc to branch so that we will try to reduce the value

B. Goldengorin and P.M. Pardalos, *Data Correcting Approaches in Combinatorial Optimization*, SpringerBriefs in Optimization, DOI 10.1007/978-1-4614-5286-7_5,
© Boris Goldengorin, Panos M. Pardalos 2012

of the current upper bound as much as possible. If this upper bound is less than the allowed accuracy, we can stop the algorithm here. This bound computation [using expression (1.3)] also grants us an insight into the arc that is most likely to cause infeasibility of the AP solution for the ATSP, so that we can use that arc as our branching variable. Thus the DC algorithm, using both necessary and sufficient conditions for optimality, extracts much more information out of the lower bound computation than any B&B algorithm, which only uses necessary conditions, and normally leads to smaller solution trees.

DC algorithms stand to gain from good lower bounds, which help to prune the solution trees. Such lower bounds allow us to discard partial solutions that are manifestly suboptimal.

Efficient implementations of DC algorithms depend on the construction of branching rules based on specific properties of the original NP-hard problem, the choice of the class of the polynomially solvable special case, and the current optimal solution to the special case. Two different approaches for creating polynomially solvable special cases are known in the literature, namely algorithmic and analytic. In the algorithmic approach, heuristic for solving the NP-hard problem is chosen, and sufficient conditions for this heuristic to return an optimal solution is formulated. In the analytic approach, sufficient conditions are given on the class of instances such that any instance can be solved to optimality, and recognized in polynomial time.

DC algorithms are designed for various classes of NP-hard problems including the quadratic cost partition (QCP), simple plant location (SPL), and Traveling Salesman problems based on the algorithmically defined polynomially solvable special cases. Results of computational experiments on the publicly available benchmark instances as well as on random instances are presented. The striking computational result is the ability of DC algorithms to find exact solutions for many relatively difficult instances within fractions of a second. For example, an exact global optimum of the QCP problem with 80 vertices and 100 % density was found within 0.22 s on a PC with 133 MHz processor, and for the SPL problem with 200 sites and 200 clients within 0.2 s on a PC with 733 MHz processor.

An interesting direction of research is to develop DC algorithms based on analytically defined polynomially solvable special cases. We plan to experiment with DC algorithms for the SPL problem based on the concept of equivalent instances (see [1]). An interesting research direction is the formulation of computationally efficient branching rules based on the properties of upper and lower tolerances for different classes of combinatorial optimization problems and their polynomially solvable special cases (see [70]).

References

1. B. Albdaiwi, B. Goldengorin, G. Sierksma, Equivalent instances of the simple plant location problem. Comput. Math. Appl. **57**(5), 812–820 (2009)
2. B.F. Albdaiwi, D. Ghosh, B. Goldengorin, Data aggregation for p-median problems. J. Comb. Optim. **21**(3), 348–363 (2011)
3. A. Alcouffe, G. Muratet, Optimal location of plants. Manag. Sci. **23**(3), 267–274 (1976)
4. Dj. A. Babayev, Comments on the note of frieze. Math. Program. **7**, 249–252 (1974)
5. E. Balas, M.W. Padberg, On the set covering problem. Oper. Res. **20**, 1152–1161 (1972)
6. E. Balas, P. Toth, Branch and bound methods, in *The Traveling Salesman Problem*, ed. by E.L. Lawler, J.K. Lenstra, A.G.H. Rinnooy Kan, D.B. Shmoys (Wiley, Chichester, 1985), pp. 361–401 (Chap. 10)
7. F. Barahona, M. Grötschel, M. Jünger, G. Reinelt, An application of combinatorial optimization to statistical physics and circuit layout design. Oper. Res. **36**(3), 493–512 (1988)
8. F. Barahona, M. Jünger, G. Reinelt, Experiments in quadratic 0–1 programming. Math. Program. **44**, 127–137 (1989)
9. J.E. Beasley. Lagrangean heuristics for location problems. Eur. J. Oper. Res. **65**, 383–399 (1993)
10. J.E. Beasley, OR-Library, http://people.brunel.ac.uk/~mastjjb/jeb/info.html. Accessed March 2012
11. S. Benati, The maximum capture problem with heterogeneous customers. Comput. Oper. Res. **26**(14), 1351–1367 (1999)
12. S. Benati, An Improved Branch & Bound Method for the Uncapacitated Competitive Location Problem. Annals. Oper. Res. **122**(1–4), 43–58 (2003)
13. V.L. Beresnev, On a problem of mathematical standardization theory. Upravliajemyje Sist. **11**, 43–54 (1973) (in Russian)
14. V.L. Beresnev, E. Kh. Gimadi, V.T. Dementyev, *Extremal Standardization Problems* (Nauka, Novosibirsk, 1978) (in Russian)
15. O. Bilde, J. Krarup, Sharp lower bounds and efficient algorithms for the simple plant location problem. Ann. Discr. Math. **1**, 79–97 (1977)
16. A. Billionet, A. Sutter, Minimization of quadratic pseudo-Boolean function. Eur. J. Oper. Res. **78**, 106–115 (1994)
17. A. Billionet, S. Elloumi, Using a Mixed Integer Programming Solver for the Unconstrained Qaudratic 0–1 Problem. Math. Progr. **109**, 55–68, (2007)
18. T.B. Boffey, *Graph Theory in Operations Research* (Academic Press, London, UK, 1982)
19. Boolean Models and Methods in Mathematics, Computer Science, and Engineering (Yves Crama, Peter L. Hammer, eds.), Cambridge University Press, Cambridge (2010)

B. Goldengorin and P.M. Pardalos, *Data Correcting Approaches in Combinatorial Optimization*, SpringerBriefs in Optimization, DOI 10.1007/978-1-4614-5286-7,
© Boris Goldengorin, Panos M. Pardalos 2012

20. Boolean Functions: Theory, Algorithms, and Applications (Yves Crama, Peter L. Hammer, eds.), Cambridge University Press, Cambridge (2011)
21. E. Boros, P. Hammer, Cut-polytopes, Boolean quadratic polytopes and nonnegative quadratic pseudo-Boolean functions. Math. Oper. Res. **18**(1), 245–253 (1993)
22. R.E. Burkard, V.G. Deineko, R. van Dal, J.A.A. van der Veen, G.J. Woeginger, Well-solvable special cases of the traveling salesman problem; a survey. SIAM Rev. **40**(3), 496–546 (1998)
23. C. Canel, B.M. Khumawala, J. Law, A. Loh, An algorithm for the capacitated, multi-commodity, multi-period facility location problem. Comput. Oper. Res. **28**, 411–427 (2001)
24. M.W. Carter, The indefinite zero-one quadratic problem. Discr. Appl. Math. **7**, 23–44 (1984)
25. R. Carragan, P. Pardalos, An exact algorithm for the maximum clique problem. Oper. Res. Lett. **9**, 375–382 (1990)
26. P. Chardaire, A. Sutter, A decomposition method for quadratic zero-one programming. Manag. Sci. **41**(4), 704–712 (1995)
27. V.P. Cherenin, Mechanization of calculations for drawing up train gathering plans. Tekhnika Zheleznykh Dorog **1**, 22–34 (1954) (in Russian)
28. V.P. Cherenin, Drawing up an optimal plan of gathering one-group trains using computers. Vestnik VNII Zheleznodorozhnogo Transp. **1**, 21–24 (1961)
29. V.P. Cherenin, Solving some combinatorial problems of optimal planning by the method of successive calculations, in *Proceedings of the Conference on Experiences and Perspectives of the Application of Mathematical Methods and Electronic Computers in Planning*. Mimeograph, Novosibirsk, 1962 (in Russian)
30. D.C. Cho, E.L. Johnson, M.W. Padberg, M.R. Rao, On the uncapacitated plant location problem. I. Valid inequalities and facets. Math. Oper. Res. **8**, 579–589 (1983)
31. D.C. Cho, M.W. Padberg, M.R. Rao, On the uncapacitated plant location problem. II. Facets and lifting theorems. Math. Oper. Res. **8**, 590–612 (1983)
32. N. Christofides, *Graph Theory: An Algorithmic Approach* (Academic, London, 1975)
33. G. Cornuejols, M.L. Fisher, G.L. Nemhauser, On the uncapacitated location problem. Ann. Discr. Math. **1**, 163–177 (1977)
34. G. Cornuejols, M.L. Fisher, G.L. Nemhauser, Location of bank accounts to optimize float: an analytic study of exact and approximate algorithms. Manag. Sci. **23**, 789–810 (1977)
35. G. Cornuejols, G.L. Nemhauser, L.A. Wolsey, The uncapacitated facility location problem, in *Discrete Location Theory*, ed. by R.L. Francis, P.B. Mirchandani (Wiley, New York, 1990) (Chap. 3)
36. G. Cornuejols, J.M. Thizy, A primal approach to the simple plant location problem. SIAM J. Algebraic Discr. Meth. **3**, 504–510 (1982)
37. W.J. Cook, W.H. Cunningham, W.R. Polleyblank, A. Schrijver, *Combinatorial Optimization* (Wiley, New York, 1998)
38. P.M. Dearing, P.L. Hammer, B. Simeone, Boolean and graph theoretic formulations of the simple plant location problem. Transp. Sci. **26**, 138–148 (1992)
39. R. Dechter, J. Pearl, Generalized best-first search strategies and the optimality of A^*. J. ACM **32**, 505–536 (1985)
40. M. Dell'Amico, F. Maffioli, S. Martello (eds.), *Annotated Bibliographies in Combinatorial Optimization* (Wiley, Chichester, 1997)
41. A. Dolan, J. Aldous, *Networks and Algorithms. An Introductory Approach* (Wiley, Chichester, 1993)
42. J. Edmonds, Matroids, submodular functions, and certain polyhedra, in *Combinatorial Structures and Their Applications*, ed. by R.K. Guy, H. Hanani, N. Sauer, J. Schönheim (Gordon and Beach, New York, 1970), pp. 69–87
43. M.A. Efroymson, T.L. Ray, A branch-bound algorithm for plant location. Oper. Res. **14**, 361–368 (1966)
44. D. Erlenkotter, A dual-based procedure for uncapacitated facility location. Oper. Res. **26**, 992–1009 (1978)

45. A. Frank, Matroids and submodular functions, in *Annotated Bibliographies in Combinatorial Optimization*, ed. by M. Dell'Amico, F. Maffiolli, S. Martello (Wiley, New York, 1997), pp. 65–80

46. A.M. Frieze, A cost function property for plant location problems. Math. Program. **7**, 245–248 (1974)

47. R.D. Galvão, L.A. Raggi, A method for solving to optimality uncapacitated location problems. Ann. Oper. Res. **18**, 225–244 (1989)

48. M.R. Garey, D.S. Johnson, *Computers and Intractability, a Guide to the Theory of NP-Completeness* (Freeman, San Francisco, 1979)

49. R.S. Garfinkel, A.W. Neebe, M.R. Rao, An algorithm for the M-median plant location problem. Transp. Sci. **8**, 217–236 (1974)

50. A.V. Genkin, I.B. Muchnik, Optimum algorithm for maximization of submodular functions. Autom. Remote Contr. USSR **51**, 1121–1128 (1990)

51. R. Germs, B. Goldengorin, M. Turkensteen, Lower tolerance-based branch and bound algorithms for the ATSP. Comput. Oper. Res. **39**(2), 291–298 (2012)

52. P.C. Gilmore, E.L. Lawler, D.B. Shmoys, Well-solved special cases, in *The Traveling Salesman Problem*, ed. by E.L. Lawler, J.K. Lenstra, A.G.H. Rinnooy Kan, D.B. Shmoys (Wiley, Chichester, 1985), pp. 87–143 (Chap. 4)

53. D. Ghosh, B. Goldengorin, G. Sierksma, Data correcting algorithms in combinatorial optimization, in *The Handbook of Combinatorial Optimization*, suppl. vol. B, ed. by D.-Z. Du, P.M. Pardalos (Springer, Berlin, 2005), pp. 1–53

54. D. Ghosh, B. Goldengorin, G. Sierksma, Data correcting: a methodology for obtaining near-optimal solutions, in *Operations Research with Economic and Industrial Applications: Emerging Trends*, ed. by S.R. Mohan, S.K. Neogy (Anamaya, New Delhi, 2005), pp. 119–127 (Chap. 9)

55. F. Glover, G.A. Kochenberger, B. Alidaee, Adaptive memory tabu search for binary quadratic programs. Manag. Sci. **44**(3), 336–345 (1998)

56. B.I. Goldengorin, The design of optimal assortment for the vacuum diffusion welding sets. Standarty i Kachestvo **2**, 19–21 (1975) (in Russian)

57. B. Goldengorin, Methods of solving multidimensional unification problems. Upravljaemye Sist. **16**, 63–72 (1977) (in Russian)

58. B.I. Goldengorin, *On the Stability of Solutions in Problems with a Single Component of Local Minima. Models and Methods of Solving Problems of Economic Systems Interaction* (Nauka, Novosibirsk, 1982), pp. 149–160 (in Russian)

59. B.I. Goldengorin, A correcting algorithm for solving some discrete optimization problems. Soviet Math. Dokl. **27**, 620–623 (1983)

60. B. Goldengorin, A correcting algorithm for solving allocation type problems. Autom. Remote Contr. **45**, 590–598 (1984)

61. B. Goldengorin, Correcting algorithms for solving multivariate unification problems. Soviet J. Comput. Syst. Sci. **1**, 99–103 (1985)

62. B.I. Goldengorin, A decomposition algorithm for the unification problem and new polynomially solvable cases. Soviet Math. Dokl. **33**, 578–583 (1986)

63. B.I. Goldengorin, On the exact solution of unification problems by correcting algorithms. Soviet Phys. Dokl. **32**, 432–434 (1987)

64. B. Goldengorin, *Requirements of Standards: Optimization Models and Algorithms* (Russian Operations Research, Hoogezand, 1995)

65. B. Goldengorin, G. Gutin, Polynomially solvable cases of the supermodular set function minimization problem. Research Report TR/6/98, Department of Mathematics and Statistics, Brunel University, Uxbridge, Middlesex, 1998, p. 5

66. B. Goldengorin, G. Sierksma, G.A. Tijssen, M. Tso, The data-correcting algorithm for minimization of supermodular functions. Manag. Sci. **45**, 1539–1551 (1999)

67. B. Goldengorin, D. Ghosh, G. Sierksma, Improving the efficiency of branch and bound algorithms for the simple plant location problem. Lect. Notes Comput. Sci. **2141**, 106–117 (2001)

68. B. Goldengorin, D. Ghosh, G. Sierksma, Branch and peg algorithms for the simple plant location problem. Comput. Oper. Res. **30**, 967–981 (2003)
69. B. Goldengorin, G.A. Tijssen, D. Ghosh, G. Sierksma, Solving the simple plant location problem using a data correcting approach. J. Global Optim. **25**(4), 377–406 (2003)
70. B. Goldengorin, G. Sierksma, Combinatorial optimization tolerances calculated in linear time. SOM Research Report 03A30, University of Groningen, Groningen, 2003, p. 6
71. B. Goldengorin, G. Sierksma, M. Turkensteen, Tolerance based algorithms for the ATSP. Lect. Notes Comput. Sci. **3353**, 222–234 (2005)
72. B. Goldengorin, D. Ghosh, The multilevel search algorithm for the maximization of submodular functions applied to the quadratic cost partition problem. J. Global Optim. **32**(1), 65–82 (2005)
73. B. Goldengorin, Maximization of submodular functions: theory and enumeration algorithms. Eur. J. Oper. Res. **198**, 102–112 (2009)
74. B. Goldengorin, D. Krushinsky, Complexity evaluation of benchmark instances for the p-median problem. Math. Comput. Model. **53**(9–10), 1719–1736 (2011)
75. B. Goldengorin, D. Krushinsky, A computational study of the pseudo-Boolean approach to the p-median problem applied to cell formation. Lect. Notes Comput. Sci. **6701**, 503–516 (2011)
76. R.P. Grimaldi, *Discrete and Combinatorial Mathematics: An Applied Introduction*, 3rd edn. (Addison-Wesley, New York, 1994)
77. M. Guignard, K. Spielberg, Algorithms for exploiting the structure of the simple plant location problem. Ann. Discr. Math. **1**, 247–271 (1977)
78. G. Gutin, A.P. Punnen, *The Traveling Salesman Problem and Its Variations* (Kluwer Academic, Dordrecht, 2002)
79. P. Hammer, Some network flow problems solved with pseudo-Boolean programming. Oper. Res. **13**, 388–399 (1965)
80. P.L. Hammer, Plant location—a pseudo-Boolean approach. Isr. J. Technol. **6**, 330–332 (1968)
81. P. Hammer, S. Rudeanu, *Boolean Methods in Operations Research and Related Areas* (Springer, Berlin, 1968)
82. J. Håstad, Some optimal in-approximability results, in *Proceedings 29th Annual ACM Symposium on Theory of Computation*, (El Paso, Tex., May 4–6). ACM, New York, 1–10 (1997)
83. M. Held, P. Wolfe, H.P. Crowder, Validation of subgradient optimization. Math. Program. **6**, 62–88 (1974)
84. P.C. Jones, T.J. Lowe, G. Muller, N. Xu, Y. Ye, J.L. Zydiak, Specially structured uncapacitated facility location problems. Oper. Res. **43**, 661–669 (1995)
85. S.N. Kabadi, Polynomially solvable cases of the TSP, in *The Traveling Salesman Problem and Its Variations*, ed. by G. Gutin, A.P. Punnen (Kluwer Academic, Dordrecht, 2002), pp. 489–583 (Chap. 11)
86. F. Kalantari, A. Bagchi, An algorithm for quadratic zero one. Technical Report LCSR-TR-112, Department of Computer Science, Rutgers University, 1988
87. R.M. Karp, Reducibility among combinatorial problems, in *Complexity of Computer Computations*, ed. by R.E. Miller, J.W. Thatcher (Plenum, New York, 1972), pp. 85–103
88. R.M. Karp, A patching algorithm for the nonsymmetric traveling salesman problem. SIAM J. Comput. **8**(4), 561–573 (1979)
89. V.R. Khachaturov, Some problems of the successive calculation method and its applications to location problems. Ph.D Thesis, Central Economics & Mathematics Institute, Russian Academy of Sciences, Moscow, 1968 (in Russian)
90. V.R. Khachaturov, *Mathematical Methods of Regional Programming* (Nauka, Moscow, 1989) (in Russian)
91. B.M. Khumawala, An efficient branch and bound algorithm for the warehouse location problem. Manag. Sci. **18**, B718–B731 (1972)
92. C.-W. Ko, J. Lee, M. Queyranne, An exact algorithm for maximum entropy sampling. Oper. Res. **43**, 684–691 (1995)

93. M. Körkel, On the exact solution of large-scale simple plant location problems. Eur. J. Oper. Res. **39**, 157–173 (1989)

94. B. Korte, J. Vygen, *Combinatorial Optimization. Theory and Algorithms* (Springer, Berlin, 2000)

95. J. Krarup, P.M. Pruzan, The simple plant location problem: a survey and synthesis. Eur. J. Oper. Res. **12**, 36–81 (1983)

96. A. Krause, SFO: a toolbox for submodular function optimization. J. Mach. Learn. Res. **11**, 1141–1144 (2010)

97. H.W. Kuhn, The Hungarian method for the assignment problem. Naval Res. Logist. Q. **2**, 83–97 (1955)

98. M. Labbé, F.V. Louveaux, Location problems, in *Annotated Bibliographies in Combinatorial Optimization*, ed. by M. Dell'Amico, F. Maffioli, S. Martello (Wiley, Chichester, 1997), pp. 264–271 (Chap. 16)

99. M. Laurent, Max-cut problem, in *Annotated Bibliographies in Combinatorial Optimization*, ed. by M. Dell'Amico M, F. Maffioli, S. Martello (Wiley, Chichester, 1997), pp. 241–259 (Chap. 15)

100. E.L. Lawler, J.K. Lenstra, A.G.H. Rinnooy Kan, D.B. Shmoys, *The Traveling Salesman Problem* (Wiley, Chichester, 1985)

101. J. Lee, Constrained maximum-entropy sampling. Oper. Res. **46**, 655–664 (1998)

102. H. Lee, G.L. Nemhauser, Y. Wang, Maximizing a submodular function by integer programming: polyhedral results for the quadratic case. Eur. J. Oper. Res. **94**, 154–166 (1996)

103. L. Lovasz, Submodular functions and convexity, in *Mathematical Programming: The State of the Art*, ed. by A. Bachem, M. Grötschel, B. Korte (Springer, Berlin, 1983), pp. 235–257

104. C.K. Martin, L. Scharge, Subset coefficient reduction cuts for 0–1 mixed integer programming. Oper. Res. **33**, 505–526 (1985)

105. M. Minoux, Accelerated greedy algorithms for maximizing submodular set functions, in *Actes Congres IFIP*, ed. by J. Stoer (Springer, Berlin, 1977), pp. 234–243

106. J.G. Morris, On the extent to which certain fixed charge depot location problems can be solved by LP. J. Oper. Res. Soc. **29**, 71–76 (1978)

107. C. Mukendi, Sur l'Implantation d'Èquipement dans un Reseu: Le Problème de m-Centre. Thesis, University of Grenoble, 1975

108. K.G. Murty, An algorithm for ranking all the assignments in order of increasing cost. Oper. Res. **16**, 682–687 (1968)

109. G.L. Nemhauser, L.A. Wolsey, M.L. Fisher, An analysis of approximations for maximizing submodular set functions—I. Math. Program. **14**, 265–294 (1978)

110. G.L. Nemhauser, L.A. Wolsey, Maximization submodular set functions: formulations and analysis of algorithms, in *Studies on Graphs and Discrete Programming*, ed. by P. Hansen (North-Holland, Amsterdam, 1981), pp. 279–301

111. G.L. Nemhauser, L.A. Wolsey, *Integer and Combinatorial Optimization* (Wiley, New York, 1988)

112. E.D. Nering, A.W. Tucker, *Linear Programs and Related Problems* (Academic, San Diego, 1993)

113. M. Padberg, The Boolean quadratic polytope: some characteristics, facets and relatives. Math. Program. **45**, 139–172 (1989)

114. P.M. Pardalos, G.P. Rodgers, Computational aspects of a branch and bound algorithm for quadratic zero-one programming. Computing **40**, 131–144 (1990)

115. P.M. Pardalos, F. Rendl, H. Wolkowicz, The quadratic assignment problem: a survey and recent developments. DIMACS Ser. Discr. Math. Theor. Comput. Sci. **16**, 1–42 (1994)

116. D.W. Pentico, The assortment problem with nonlinear cost functions. Oper. Res. **24**, 1129–1142 (1976)

117. D.W. Pentico, The discrete two-dimensional assortment problem. Oper. Res. **36**, 324–332 (1988)

118. A. Petrov, V. Cherenin, An improvement of train gathering plans design methods. Zheleznodorozhnyi Transp. **3**, 60–71 (1948) (in Russian)

119. S. Poljak, F. Rendl, Node and edge relaxations of the max-cut problem. Computing **52**, 123–137 (1994)
120. S. Poljak, F. Rendl, Solving the max-cut problem using eigenvalues. Discr. Appl. Math. **62**, 249–278 (1995)
121. G. Reinelt, TSPLIB 95 (1995), http://comopt.ifi.uni-heidelberg.de/software/TSPLIB95/. Accessed April 2012
122. C.S. ReVelle, H.A. Eiselt, M.S. Daskin, A bibliography for some fundamental problem categories in discrete location science. Eur. J. Oper. Res. **184**, 817–848 (2008)
123. C.S. Revelle, G. Laporte, The plant location problem: new models and research prospects. Oper. Res. **44**, 864–874 (1996)
124. F. Rendl, G. Rinaldi, A. Wiegele, Solving Max-Cut to optimality by intersecting semidefinite and polyhedral relaxations. Math. Progr. **121**, 307–335, (2010)
125. T.G. Robertazzi, S.C. Schwartz, An accelerated sequential algorithm for producing D-optimal designs. SIAM J. Sci. Stat. Comput. **10**, 341–358 (1989)
126. L. Schrage, Implicit representation of variable upper bounds in linear programming. Math. Program. Study **4**, 118–132 (1975)
127. A. Schrijver, A combinatorial algorithm minimizing submodular functions in strongly polynomial time. J. Comb. Theor. Ser. B **80**(2), 346–355 (2000)
128. A. Tripathy, H. Süral, Y. Gerchak, Multidimensional assortment problem with an application. Networks **33**, 239–245 (1999)
129. V.A. Trubin, On a method of solution of integer programming problems of a special kind. Soviet Math. Dokl. **10**, 1544–1546 (1969)
130. M. Turkensteen, D. Ghosh, B.Goldengorin, G. Sierksma, Tolerance-based branch and bound algorithms for the ATSP. Eur. J. Oper. Res. **189**, 775–788 (2008)
131. E.S. van der Poort, Aspects of sensitivity analysis for the traveling salesman problem. Ph.D. Thesis, SOM Graduate School, University of Groningen (Labyrint Publication, Capelle a/d IJssel, 1997)
132. T.J. Van Roy, L.A. Wolsey, Valid inequalities for mixed 0–1 programs. Discr. Appl. Math. **14**, 199–213 (1986)
133. V.E. Veselovsky, Some algorithms for solution of a large-scale allocation problem. Ekonomika i Matematicheskije Metody **12**, 732–737 (1977) (in Russian)
134. A.C. Williams, Quadratic 0–1 programming using the roof dual with computational results. RUTCOR Research Report 8-85, Rutgers University, 1985
135. D.P. Williamson, *Lecture Notes on Approximation Algorithms*. IBM Research Report (IBM Research Division, T.J. Watson Research Center, Yorktown Heights, Spring 1998)
136. W. Zhang, R.E. Korf, Performance of linear-space search algorithms. Artif. Intell. **79**, 241–292 (1995)